MYSTERIES
of the
ROUND TOWERS

T0246545

"*Mysteries of the Round Towers* offers a groundbreaking exploration of Ireland's enigmatic stone structures, revealing their subtle magnetic and energetic properties. Through a form of dowsing, Freeland senses an ancient wisdom behind these stone towers and provides new insights into their purpose. Modern scientific studies have shown that humans can sense exquisitely sensitive magnetic fields, so while dowsing may once have been considered folklore, science is slowly catching up with these traditional means of sensing the environment. Researchers like Freeland are at the forefront of using these techniques to rediscover what the ancients undoubtedly knew."

DEAN RADIN, PH.D., CHIEF SCIENTIST AT THE
INSTITUTE OF NOETIC SCIENCES (IONS)

"With their distinctive double-walled tower design, strange doorways, and unique cyclopean construction, it would seem that it would take a historical detective—or someone well versed in the nature and use of stone—to solve this mystery. In *Mysteries of the Round Towers*, Christopher Freeland does just that. The only other structures comparable to Ireland's round towers are ancient monolithic obelisks found around the world. They have one thing in common with the towers of Ireland: they have the same magnetic disposition, an important clue to their purpose. Freeland makes a compelling case that the subtle, chthonic energies within the earth were part of a now lost sacred science understood by our ancient ancestors. Freeland's knowledge, personal experience, and research are impeccable. And the spirit in which he writes

Mysteries of the Round Towers makes it interesting and informative as well as a pleasure to read."

"Christopher Freeland has shown that those who built the round towers had a deep understanding—an understanding we have lost in a materialistic, scientific culture—of the effects of energetic forces and ways to construct conduits to conduct their energies for health. The towers probably continue to be operable—why not?—so they would send energies to those around them, though how far we can't guess. We remain in the thrall of domains we can't see, both modern and ancient. The towers speak to the mystery of unseen forces that drive human behavior and seek to rectify human damage in powerful ways."

MYSTERIES
of the
ROUND TOWERS

The Subtle Energies of the
Stone Structures of Ireland

A Sacred Planet Book

CHRISTOPHER FREELAND

Bear & Company
Rochester, Vermont

Bear & Company
One Park Street
Rochester, Vermont 05767
www.BearandCompanyBooks.com

Bear & Company is a division of Inner Traditions International

Sacred Planet Books are curated by Richard Grossinger, Inner Traditions editorial board member and cofounder and former publisher of North Atlantic Books. The Sacred Planet collection, published under the umbrella of the Inner Traditions family of imprints, includes works on the themes of consciousness, cosmology, alternative medicine, dreams, climate, permaculture, alchemy, shamanic studies, oracles, astrology, crystals, hyperobjects, locutions, and subtle bodies.

Cataloging-in-Publication Data for this title is available from the Library of Congress

ISBN 978-1-59143-528-0 (print)
ISBN 978-1-59143-529-7 (ebook)

Printed and bound in the United States by Lake Book Manufacturing, LLC

10 9 8 7 6 5 4 3 2 1

Text design and layout by Virginia Scott Bowman
This book was typeset in Garamond Premier Pro, Myriad Pro, and Futura with Begum used as the display typeface

To send correspondence to the author of this book, mail a first-class letter to the author c/o Inner Traditions • Bear & Company, One Park Street, Rochester, VT 05767, and we will forward the communication, or contact the author directly at **radiesthesia.online**.

■ ■ ■

Dedicated to those industrious masons who toiled on the round towers and sacrificed their well-being for the benefit of their fellows and future generations.

Contents

Foreword

Jeremy Massey

Myriad attempts have been made to grapple with the mystery of the round towers of Ireland over the many hundreds of years since their construction, and while some have come maddeningly close to making sense of them, their true purpose has eluded every mind that's ever been bold enough to scrutinize them. Until, perhaps, now.

Using the humble and profoundly accurate method of radiesthesia and ably assisted by his nimble intellect and tireless curiosity, Christopher Freeland has managed to pierce the enigma of the ancient towers and deliver to us a sound understanding of their raison d'être. The official dogma that it was the church that built the towers is deftly dismantled by Freeland's theory, which can be demonstrated through active experimentation.

Standing on the shoulders of the ancient stonemasons, who had a firm understanding of the flow of energy in the earth, Freeland helps us come to grips with Nature's alchemy of stone, water, and magnetism, one of the fundamental forces at work on our earth, and the connection the towers have to them, ultimately unmasking the secret of the arcane wisdom held within.

In his erudite and elegant prose, Freeland brings us through many elucidating moments of his own discoveries and experiments with radiesthesia, frequency, and magnetism to deliver a persuasive and gratifying conclusion that will not only alter the way you look at the round towers of Ireland but the way you look at Nature itself.

More than a singular work, this little book is a polished gem of concentrated wisdom from a man whose findings have been collected on the empirical path by an open and curious mind.

JEREMY MASSEY is a man of letters and filmmaker as well as third-generation undertaker who worked with his father for many years at the family firm in Dublin. A screenwriter by training and author of *The Last Four Days of Paddy Buckley*, Massey lives in Ireland.

The Research Method

The development of any culture is directly related to the understanding of its environment.

CALLUM COATES

H olistically, the subject of this book is stone but with a distinct emphasis on the application unique to the round towers of Ireland and Scotland, for they incorporate attributes which, I venture, are explored here for perhaps the first time in human history. (My presumptuous belief knows no bounds!)

Stone, along with water, is the essence of the earth, and although it has been much appreciated for many of its properties since time immemorial, it is not exactly the sort of thing that excites interest, let alone fires the imagination. So I am hoping that by bringing to light a few realities on this superficially mundane substance I will provide insight into some of its inherent properties not fully appreciated in the collective conscious—specifically, its magnetic capacity, its close relationship with water, and the gracious benefit it provides. If we neglect—as we systematically do—such key characteristics, we may be doomed to remain in nescience and miss out on some of the advantages that our ancestors were fully aware of and generously bequeathed to us. Indeed, if we neglect to investigate any aspect of what we are able to observe and comprehend, I believe that in all probability we are destined to regurgitate what others have said, simply repeating their narrative and, sadly, never becoming any the wiser.

This book is a small contribution to what deserves to be a full-blown investigation into the exact purpose of the round towers and is based on a few items of evidence that have, for one reason or another, escaped the attention of other observers. These truths will not be to the taste of many, first because they fall totally outside of the mainstream story, secondly, because consideration is given to some essential but neglected factors in the affairs of architecture, and thirdly, because those factors by their very nature do not fall neatly into what is commonly accepted in the West as the scientific paradigm.

What is more, the research method used to determine these special characteristics—namely, radiesthesia, a.k.a. divination, water-witching, or dowsing—is unorthodox, even described by some of its detractors as pseudoscientific. Having said that, I would ask that you at least hear me out and then after careful consideration, draw your own conclusions, and even better, start your own inquiry into the evidence we have at our disposal so as to understand what could have possibly happened to eclipse this science—for that is what it must have been for a fairly lengthy period in human history. Finally, for the curious and the practical-minded reader, this book hints at some of the obscure techniques used by the ancients and provides a look at the possible applications in our modern age.

I believe that up until this present moment in time, the facts presented here have been universally ignored for the most part, except perchance among water-divining circles. Therefore, they may require a period of reflection and experimentation before they gain currency, or rather, establish a renewal of ancient tradition.

Within this book, there is no desire to impose a belief, for we are highly unlikely to know with precision what was in the minds of the beings who built these singular stone structures, which are still with us after many hundreds of years, in fine fettle but unfortunately not in prime working order. And neither is there is a specific intention to disprove other points of view, although that may be the result, given the illogical reasoning expressed by the protagonists themselves in their haste to reiterate the accepted paradigm and garner further acceptance from the herd.

However, as the legal maxim would have it, "No one is bound to give information about things he is ignorant of, but everyone is bound to know that which he gives information about." Therefore, every effort has been taken to include as much of the data and literature concerning the round towers as possible, so please excuse my shortcomings and what might be seen as a contrarian attitude.

Keep in mind, though, that a holistic method of inquiry cannot exclude observation of what is actually present, irrespective of whether it meets mainstream approval or not. One thing is for sure, if we blindly repeat what others say and close our minds to relevant facts, intolerance and prejudice will be close at hand. What follows in the upcoming sections and chapters is a shot at describing what the significance of these facts might be.

DIVINATION AS A RESEARCH METHOD

You might agree that for information to be considered worthwhile and useful, it must be essential, practical, and authentic. Essential because it benefits one and all, serving specific objectives; practical because its application is simple and accessible; and authentic because it is the truth. A possible definition of this expression would be "any knowledge that reaches our awareness, thereby encouraging and reinforcing a sense of unicity; a practical, comprehensive gnosis that remains constant within the bounds of natural behavior and events."

Divination has always provided humans with the ability to access information of quality. The acid test is naturally one of value—we are human after all—and of course values change over time, as a function of a host of factors. The most powerful influence over human values would seem to be emotional security—in other words, feeling safe. Our instincts operate in that domain as we strive for food, shelter, reproduction, sleep, and security, all of which provide a sense of well-being once achieved. If one takes Nature as the guide, for she is the ruler, then what is harmonious wins the day, closely followed by what is useful and practical, so long as the aim is one of comprehensive, all-embracing harmony. Our instincts follow this path and have throughout history.

However, we moderns have so much information at our disposal—increasingly so of the superfluous variety—it is quite hard to discriminate what is good and bad for us. Particularly when "good" and "bad" are no longer subjects for us to decide upon, as the practical objectives have been turned to commercial gain by means of manipulation, then fed back to us to consume. The danger is that we throw the baby out with the bath water as we either focus on the objects of desire and greed or simply give up to the overweening manipulation and stop thinking.

Our sources of information deserve a closer look to see if we can surprise an insight as to how we actually acquire knowledge. If life is movement (as opposed to stasis), and movement is a function of the various forms of energy coming in from the cosmos and the resultant interaction here on the surface of the earth, it might be wise to consider observation of these movements as a source of understanding, *if* we are able to associate an event with a result. Regrettably, over the course of time, this ability to observe and synthesize the whole became a very binary affair, and we ended up with dogmatic cause and effect, the most basic misconception imaginable because it lacks the subtle exchange offered by the holistic approach, where everything has its place, its cycle of life, and its demise.

There are two very ancient methods of divining or accessing information—astrology and radiesthesia. Both can be freely obtained with some effort, but as is so often the case with something that is free of charge, we look askance and tend to diminish the value of these methods, even disdaining and rejecting what they offer. Yet, we would do well to recall that Nature provides us with our very life for free; we have only to put some effort into maintaining it. Having said that, by our modern standards, neither of those techniques are "free," they both require a great deal of time and effort in order to become proficient, although radiesthesia is the hands-down winner as results can be immediate. Confidence, however, in the exactitude of the answers, is a question of practice, and probably of years.

These two methods have been practiced both openly and occultly since time immemorial, in all civilizations and cultures. They are what I call scientific arts, or "sci-arts," par excellence, and two of the best ways to discover what is going on and what is about to happen. When these methods are in the hands of talented practitioners, there are no secrets.

RADIESTHESIA

A deep dive into astrology is beyond the scope of this study, since most directly related to the details uncovered about the towers is the other form of divination. Not as familiar as the planetary art, radiesthesia is the name given to dowsing by a French group whose proponents tried hard to get dowsing accepted by the scientific community as a valid means of research and analysis in the field of physics. Unfortunately perhaps for the advancement of the field, they were unsuccessful. An image may spring to mind here: the guy with the hazel twig looking for water; the soldier with the rods looking for tunnels; or yet again, a person sitting at their desk, pendulum in hand, bent over a map searching for mineral deposits in some distant land. Those are not just possibilities. They are actual and much like the astrologer working in a bank, deciding the right moment to invest, in what and how much.

Compared to the more general and symbolic approach of astrology, radiesthesia is a down-to-earth method that requires a questioning process between the human rational function and the fundamental source that provides the answers—the all-aware, omnipresent consciousness that is shared by all that exists, in every realm, at all times.

Radiesthesia or any other form of kinesiological muscle testing serves nicely, especially as one only needs one hand to operate when dowsing. There are no doubt other equally precise means, but few so practical for accessing truth, or quality information, instantaneously, without any need to cast a chart or consult the ephemeris or a computer. While the term *radiesthesia* was coined in 1920, the practice has been around forever. It involves a tedious training involving practice and yet more practice and a special vacuous mindset where you "know" nothing, so it does not enjoy the same popularity or renown as astrology.

The accomplished radiesthesist is able to provide an immediate answer to a question. This is not quite the case with astrology. While horary is a most useful form of astrology, it is definitely a task for the experienced practitioner, and it is highly unlikely that you would be prepared to pay $200 or so to find out the answer to a simple question of a humdrum nature, but it takes more time to prepare a chart and interpret the symbolism so is probably justified.

Radiesthesia is an "intrinsic" method that is sufficiently simple to be accessible to one and all. When I say "intrinsic," I mean that it is in opposition to the input from the planets and cosmos. The frequential energy pouring into our environment makes for a constantly changing mix, which is in all probability quite impossible for an untrained human to fathom. But once the trained astrologer can appreciate the patterns and make the correlation between cause and effect, recognizing the interaction at work among the multiple factors involved, the picture becomes clearer, and we are "in the know."

It is somewhat different with radiesthesia, as one needs to assess the context and to ask the right questions, a highly demanding task calling on a careful combination of intellectual ability and intuition, with a large dose of humility.

Both methods have encountered mixed fortunes as the ages have progressed, but ever since the age of enlightenment dawned in the eighteenth century, the eclipse of both has apparently been nigh on total. Radiesthesia made a brief comeback between 1910 and 1950, especially in France. It was extensively used by French doctors in the early 1950s as a very simple but efficient means to check if diagnosis and prescription were correct. That did not last long, but at least they were no longer burned at the stake.

The task is neither easy nor without risk of incurring enmity and wrath by upsetting the accepted paradigm. We have now become so used to being told what is good for us, even the most intelligent people accept without a murmur as soon as a prominent "name" says it is good or bad, or a household brand is determined to be just what you need. A clear sign that our discrimination has been demolished.

Radiesthesia offers the ability for the modern individual to discriminate, to determine what is real and what is beneficial amid the modern world's maelstrom of conflicting information. In our domain of multiple senses, unicity could be said to be the foremost human sense, experienced by some as the feeling of being at one, at-one-ment if you like.

That being-at-one sensation is, I believe, the result of feeling that you are in exactly the right place, which it would be reasonable enough to assume, happens when your physical body is in harmony with the surroundings. To do that, the correct balance of positive/negative, attraction/repulsion opposites must presumably be in sync; in other words, the "magnetic" component is as close to harmonious as possible. We will take a closer look at this magnetic component shortly.

Human evolution, which of course happens for our anthropocentric species, depends totally on incoming cosmic frequencies—whether we like it or not—if perchance we can recognize that truth. Unfortunately for a little over a century and a half, we have managed to interfere with those frequencies with electricity and, more recently, electromagnetic radiation, in such a way that certain categories of disease have been introduced into our environment, and the earth, along with its magnetic field and its denizens, are weakening. But of course, it is not our fault, so we do not even deign to investigate that possibility, and I suspect vested industrial interests will ensure that we will not do so in the near future.

Naturally enough, these frequencies accumulate in all elements, thanks to Nature's alchemy, which aligns the component molecules over time, as a function of the type of element and surrounding conditions, resulting in an aggregate magnetic charge that varies from one type of substance to another depending on the specific locality, different periods of day and night, and all sorts of other factors.

The problem of not recognizing the unicity of this energy and all aspects of its presence—physical/subtle, kinetic/potential—is that we fail to observe the underlying harmony that is to be found in all components that make up Nature. Observation is a forgotten, almost extinct art. We know of train spotters, bird watchers, astronomers, philatelists, and a host of other wonderful hobbyists whose leisure time is absorbed by their singular passion. But there are not many Taoists in the neighborhood who spend their lives shadowing Nature!

One thing is certain: the ability to recognize the fact that we are an integral part of Nature, belonging to this phenomenal universe as components rather than as its masters, is rare, but essential if we hope to surprise her secrets. It is this connection to all of Nature that we are tapping into when we employ radiesthesia.

THE PROCESS OF RADIESTHESIA

There is no need for anything of a philosophical, psychological, or psychic nature. There is no requirement for a belief system. Radiesthesia is a totally practical, simple method that requires, as mentioned above, a vast amount of practice, careful thought in formulating the questions, discipline, and—at some stage—a quiet conviction as to how the whole process works. This last component is of the essence, the consequence of truth becoming apparent repeatedly via answers to multiple questions that have, in time and context, been revealed in their true verifiable nature. Now there is a tool worth using!

The pendulum is surely the most flexible of all the tools that can be used to find answers to questions. It is a "barefoot" way par excellence to discover if something (food, drink, person, medication, situation) is beneficial or not for you or others. Because of its flexibility and simplicity of use, we will focus on the pendulum rather than other methods, such as a dowsing rod, the hazel twig or equivalent, or the Bi-Digital O-Ring, friction, arm muscle, sway, toe-touching tests, which while equally valid, shall not be considered here.

There is very little that cannot be determined using a pendulum, and not a single domain in our physical life where it cannot be applied, though becoming confident in its use requires a lot of work, involving not just practice, but understanding, if one is to acquire the desired intuition. However, almost everyone can become proficient in its use.

Perhaps a brief explanation as to the use—or more justly, my use—of a pendulum would add a little clarity. The French and Belgians between 1900 and the 1970s wrote prolifically on the subject of radiesthesia. Based on some of these writings, and using their training methods and techniques, I developed a structured system of using a pendulum for an increasingly broad range of applications. There is a delicate collaboration required between the cerebral intelligence and intuition, this can be developed and encouraged by practice, but of a very deliberate kind. Once one has established what is known as the Convention, where the operator decides how the pendulum should move to manifest the "yes" and "no," after all it is operating as a result of a neuromuscular impulse, and they are your nerves under your instruction, one then needs to not only become familiar with the requisite State of Mind, it is best to practice that regularly until it becomes a habit. My training as a monk in my youth, and maintained ever since, provides the ideal meditative state, and the no-mind brainwave is readily accessed. There are four stages in the mental process before handing over to the intuition, and these are:

1. Mental designation of the object searched for, mentally envisaging and then orienting the mind toward it. One needs to define quite clearly what it is one is after, in as simple and precise terms as possible.

2. Formulation in so many words of exactly what you have just defined as the object of the research. It is the expression of your aim.

3. The passive state of neutrality. Having formulated exactly the objective to be achieved, the operator's all-conscious gently shifts into alert mode, a state of calm.

4. Stating the question in one's mind and retaining focus until the answer comes.

Success in a search very much depends on two things: The strict application of the 4 Stages, and a calm state of mind during the entire search process. And presto! The pendulum moves indicating either a yes or a no, or if you are working with a chart, it indicates the answer. I use a chart when dealing with earth energies and most aspects relating to the towers, for example, "Is the tower over the crossing of underground water?" Or, "What is the magnetic polarity of this section of a stone?" asked while pointing your finger or a pointing device.

I personally use a bronze pendulum weighing about 25 grams so it is heavy enough for use outside in the wind. I hold it in my right hand and the chart in my left. I make my own charts so as to contain the relevant, detailed information. Answers come relatively quickly, rotating clockwise for the yes or positive polarity.

Most people who use a pendulum share the belief that everything that exists has its specific frequency or vibration. Now if you relate to that frequency, and why not, you are in a direct link with that object, and by definition no longer "connected" to whatever else is in the vicinity. This frequential approach is known as the physicist school of radiesthesia, very closely related to modern science, where everything is dissociated and inspected in its own light, so to speak.

The mentalist school—the other school defined by the French— works on a much more psychological principle, whereby the intellectual faculty of the mind asks a question of the subconscious component of the mind because it does not know the answer, and thanks to a neuromuscular response, a reply comes from some undefined entity, which in theory is in communication with the subconscious.

If, on the contrary, you consider yourself, as I do, to be a part of rather than apart from everything that exists, the process becomes remarkably different. At the time that you recognize that you don't know something, you are potentially open to whatever is actually there. But if you focus on a specific frequency, you again create a subject-object relationship, and that is not where you need to be to achieve an accurate response in radiesthesia. It seems reasonable to say that the best possible position to receive quality information is in a state of vacuity, with nothing imposed on the mind-space environment. There is a greater chance of connecting with the total environment (rather than the mental space and whatever is there) when you feel you are a component rather than some sort of controlling influential factor.

The danger, however, is a lack of vigilance because that vacuity can be invested by uninvited ideas or forms. Consequently, a strong sense of discrimination is essential, to be employed at all times to ensure its firm anchoring in the state of being of the radiesthesist, whereby awareness or recognition of any impression coming into the mind-space is calmly acknowledged. Such a state can be described as being open to an impression, the sense of radiesthesia, as given by the two French churchmen, Alexis Timothée Bouly (1865–1958), and Louis Bayard (1861–1950) in the early twentieth century.

Neither the mentalist nor physicist schools fit this open mindset. The two are mutually exclusive, and both fall short of the total reception of the required perceiving. What is worse, to my mind, it does not correspond to a holistic approach where inclusion is the key. Especially the inclusion of one's thought process or mind, as one of many similar components that make up the human segment of this dimension, which while highly important to us, is of little consequence to the flow of events from a cosmic viewpoint!

At some stage, and it will probably be early on, once you have acquired a certain ease in using a pendulum, you will ask, "What's going on here?" I make no claims to knowing any more than the next person. In fact, on the contrary, I am such an ignorant fellow that I have to use a pendulum to discover answers to the simplest of questions! An explanation as to how a neuromuscular response to a mentally formulated question can occur would be welcome. Or, at least a convincing theory as to its apparent working.

There is one thing common to all life: it is consistently subject to natural forces. No matter how hard one tries to work with or against those energies, they resist. They have a life force of their own in the same way that we do. One might even say they have an intelligence. Indeed, they do, but that is not what science would have us believe, for as anyone who has studied and worked with water will tell you, that is the only honest conclusion one can reach. The whole point, however, is that there is a component that we can perceive and feel, and another that we merely feel—namely, the phenomenally aware conscious component and the (rather rudely termed) unconscious, or all-conscious as I prefer to call it because it encompasses all components of our physical reality. Radiesthesia is the bridge between these two.

If one can accept such a theory, there is a substantial advantage in that the phenomenon of dowsing does not have to depend on some elaborate explanation of physics (based on differentiation) or magic, which means we no longer have to rely on some undefined, unknown and therefore mysterious factor. This approach to radiesthesia, based on communication between the conscious and the all-conscious, removes all need for any mystification, and at the same time puts mystification exactly where it should be: in the realm of egocentricity and real pseudoscience!

1

The Enigma of
the Round Towers

T he history of Ireland from the fifth century onward is so closely
tied to the Christian church that it has now become impossible
to discover the facts of what transpired before the arrival of the early
Christians, and there would be little exaggeration in saying that the
popes considered all people pawns in the power games they were wont
to play with anyone who had any clout that they might be in need of. In
view of the fact that Ireland was "given" to the English king Henry II
in 1155 by Pope Adrian, the documented history from thenceforward
is Anglocentric and focused on ecclesiastic affairs. So little interest is
given to the well-being of the inhabitants and their manner of living
in such a situation that only the wielders of power are given space in
the books, solely recording their benefits and feats accomplished, if any.
And so it would seem that the origins of the round towers fell into that
category of convenient oblivion.

The round towers, for those not familiar with them, are the subject
of a mystery which, while it has not perhaps attracted as many theories or as much ink flow as the Egyptian pyramids, is still one of those
unresolved, intriguing phenomena. Not only are we still very much in
the dark as to their purpose and who built them and when, but also

as to why they were built only in Ireland, Scotland, and supposedly in Douglas, Isle of Man.

They are singular structures that continue to confound the reasons put forward by modern writers and theorists for their presence in the Irish countryside and even in some urban areas (obviously of more recent development). Sometimes they are to be found in large open spaces where they provide a dominating view for anyone perched at the top; other times they appear in the depths of a valley. There is no apparent logic to be found there. However, given certain facts to be discussed later have never, to my knowledge, been considered either with regard to the round towers or in environmental studies, I will risk casting some light onto the enigma of the towers that would justify a reappraisal of the mainstream paradigm and more importantly, make way for further necessary research into their raison d'être.

According to credible resources, over the course of time, there have perhaps been a hundred or more towers, but in the general consensus now maintained, only sixty-three remain. In *The Towers and Temples of Ancient Ireland: Their Origin and History Discussed from a New Point of View*, the rector of Ennis, Marcus Keane, says that lists of round towers amount to one hundred and twenty and that sixty-six or so of them remain. Some of them are dubious by my restrictive criteria, which I will detail in later chapters, but that changes little when taking a larger view of their purpose, which could well have been to use the stone's magnetic properties to benefit us and the immediate environment.

The dimensions and the principles of the towers vary little, which would suggest they are the work of a coherent body, such as a guild working in full understanding of what they were doing. The skill and finesse deployed in the construction of the towers is worthy of Ireland's mythical builder Gobán Saor, or more likely, his lineage. It is unlikely that more than one hundred towers could be built in the lifespan of one man, but that is mere conjecture. If the criteria of modern guilds are anything to go by, one could project those standards backward to their predecessors as secrets retained by the guilds are only passed on to those worthy students who have demonstrated their ability over numerous years; otherwise they tend to be lost. Even though those in the building trade today rarely demonstrate the same expertise shown by the crafts-people of the guilds, the principle still remains: hold on to the good stuff and make sure it is passed on down to posterity.

Not enough is known about Gobán Saor for any constructive fac-tual evidence to be put forward. No doubt clues are to be found in the mythology, even if those stories are interwoven and cloaked in Christian myth with a strong affiliation with the legendary Saint Patrick and other missionaries.

It is a mere hypothesis to suggest that the origin of the towers was concomitant with the foundation of the monasteries. It cannot be proven for lack of dated epigraphic, literary, or any other form of evidence and serves no chronological purpose whatsoever. We should be aware, nevertheless, of the fact that the church has never actually, by means of any form of statement, encyclical or otherwise, claimed to have built the towers. It is far more subtle and pervasive than that. The church has allowed passing generations to believe it is the origin of the towers by gradually erecting ecclesiastical buildings beside them, and in some cases onto them, as in Turlough for example.

This action is far more sinister and malevolent than simply mislead-ing people, however, because it not only places a firm proprietary grasp on the towers and their environment, it also distorts the beneficial effect and energetic impact the towers have on the immediately available agri-cultural land. The church has buried numerous bodies all around and even up to the walls of the towers—a serious pollution of the immediate area of many of the towers, even if author George Lennox Barrow in his book *The Round Towers of Ireland* would have us believe that towers were sometimes built over *existing* graves.

Barrow claims that "there is no doubt that the use of lime mortar, like the principle of the arch, was unknown in Ireland before the com-ing of Christianity in the fifth century." That, like the horns on a hare, may well be a possibility, but it is debatable due to the scant evidence for or against it. Why Christianity was responsible for any architectural development in Ireland at that time, however, remains quite nebulous, especially since there was such clear sharing of skills in metalwork and crafts among the indigenous tribes of western Europe. The Celts being reputed, according to relatively recent academic theory, to be excellent jewelers, although the samples unearthed in Ireland, of course, do not prove that the land was necessarily inhabited by Celtic tribes. Knowledge of those techniques would probably have come from active exchange and travel among people who had common interests. The early Christian settlements consisted of primitive structures, sufficient to provide shelter, but the concept of permanent buildings came later, when the wherewithal to provide material and labor were forthcoming, but not until then. Were Christians the only travelers to Ireland in those times?

There is no point being drawn into any polemic, so let's stay with what we know and can discover about the towers, then as now, for that is where the truth lies; the why of the towers, not the who and how of the monasteries, churches, or stone structures added at a later date for their perverse aims.

It is relevant to note, however, that the technique used in building the towers is *never* used in any of the ecclesiastical buildings, reputedly associated with the church. What is more, the Roman church is silent on the application of this technique, as it is concerning the technology used in building the medieval cathedrals, so there is a strong case for the builders being totally dissociated one from the other.

Let's have a closer look at the towers themselves and their chief characteristics.

DESIGN CRITERIA

In his book, Barrow usefully records the type of stone that fifty-one of the towers left standing are comprised of. Quite a few of them are made with a mixture of stones; for example limestone for the main body work and granite for the window and doorway frames. The three towers in Scotland—Abernethy, Brechin and Egilsay—are made of sandstone. That information is not only of academic interest, it should also encourage us to get our thinking caps on.

Just down the road from where I live on the Sheep's Head is a fine stonecarver who has given me insight into the work and tribulations of the modern stoneworker, as well as information concerning the ins and outs of the work itself, including the hardness of stone and the tools necessary for working on the various categories of stone. I asked if he could give me a rough idea of how long it would take to dress a stone to be used in the building of a tower. He replied that an easy answer would be impossible to give because although the hardness of the stone would basically determine the time spent, the tools available would also be a key factor, as would the conditions of the local working facilities such as the water and shelter available to the stonecutters (you don't work so well huddled for hours over the stone you're working on if the wind is driving the rain into your body).

One cannot help but wonder if the builders of the towers might have had access to more sophisticated tools than we are led to believe were available in the early days. We can justly wonder what kind of tools those masons had to work the granite and sandstone. You cannot dress granite with a copper chisel, even with the strongest will in the world! And sandstone is no picnic either. It is almost as hard as granite and requires tough tools to make an impression on it.

Of the fifty-one towers in Barrow's book only five are made of slate, the easiest of stones to work. Two are made of granite, thirteen of sandstone, one of basalt, and the remaining thirty of limestone, which next to slate is relatively easy to work.

Most of the stone used to build the towers came from local sources apparently, but quarrying and transporting the stone is the easy part. Dressing a block to the approximate curve of a fifty-foot circumference requires a lot of effort, skill, and time. This creates another enigma much in line with the question concerning the tools available to the

Egyptian builders of the pyramids, but perhaps we have enough to deal with already without introducing another enticing rabbit hole.

The circumferences of the towers measure, by and large, between 45 and 55 feet. There are two walls of block stone and mortar—one external and one internal filled in with rock rubble and earth. The thickness of the two walls at the lowest point at which it can be measured ranges from 3 to 6 feet. We assume this from the stumps that remain. Doorways, windows/openings, story height, and diameters are also clearly defined. The elevation of the doorways varies substantially, from ground level, as at Scattery, to almost 26 feet high, as at Kilmacduagh.

A consideration that needs to be mentioned is the fact that none of the round towers have their original cap. At some stage they must have fallen off or been replaced or destroyed, so we are not even sure what form they took, and the reports on the repairs effected in the late 1800s remain silent as to the original state, if ever it was known. The few caps that have been found nearby, as with Aghagower, are conical in form with a slant of approximately 65 degrees, but a complete one has never been found. Consequently, it would not be unreasonable to question whether a tower ever had a cap, and if so, what the actual form of it might have been.

These shared particularities would indicate applications and concerns totally at odds with the mainstream theories purporting that the towers were destined to be belfries or used for sanctuary. How, for instance, are you going to get a bell through a doorway that is 16 feet above the ground? If your bolt-hole is 26 feet above the ground, you are going to need a ladder of some kind to climb up there. A ladder that long would not bend so that it could be pulled up through a "doorway" measuring only 5 feet high. If the ecclesiastics opted for a rope ladder, how would they have gotten the ladder down from the doorway when it was needed? I know I would hesitate to use one left constantly exposed to the elements and resting on damp stone where it would rot rapidly.

*Photo of Kilmacduagh's doorway height of almost 26 feet high
compared to my son, who is 5 feet 8 inches tall.*

It is a sound argument to postulate that the height of the doorway ensured stability as the structure would not be weakened by an opening, but there must have been a reason for placing openings at different heights when there are so many other standard features common to all towers.

Stone was used in Ireland from earliest times. In passing, it is not possible to provide precise dates as modern methods, such as carbon-14 testing, rarely used by geologists as it can only date an object younger than fifty thousand years or so, might be able to tell you the age of the stone, but not when the structure was built. Stone in Ireland was used as a building material for a number of purposes such as fortifications, tombs, beehive huts, and simpler structures—not many of which remain—but it was only, in all probability, from the time that mortar (cement) was introduced that higher, sturdy buildings such as the towers became possible. There are of course many examples of stone being used beforehand but not in the context of building.

Stones have been laid out and positioned in many different forms and therefore, one can reasonably assume, with different purposes. But nowhere else outside of Ireland and Scotland can you find (still standing) construction of nearly 100-foot towers with double walls and openings that defy all practical use of conventional openings, let alone of a building with recognized purpose. Who built the towers will probably remain a mystery, but it might not be somewhat farfetched to state that it was one tribe or another.

ORIGIN THEORIES

There are a number of very evident details that would point to the construction of the towers being the work of a guild or some form of esoteric brotherhood, somewhat like the builders of the medieval Gothic cathedrals. There are multiple facets common to all towers that support this theory, including dressed stones (consider how long it takes to work and shape a stone and then calculate how many stones are used in the construction); the height and orientation of the doors and openings, which cannot be haphazard; and the difficulty in capping the tower, which precludes it being the work of an amateur. Once again, available evidence gives us absolutely no indication as to how these buildings were orchestrated, but practical experience would indicate that such organization must have been extremely efficient and structured.

If, as so many theorists have suggested, the church was responsible for building the towers, a few very simple questions should find ready answers, but they are met with resounding silence:

- Why is the round tower part of an ecclesiastic enclosure, and what is its purpose?
- Why are graves dug around and in close proximity to the towers?
- Why are the doorways at different heights, and how do you withdraw a ladder into the building, especially one 26 feet tall such as the Kilmacduagh tower?
- Why hasn't a bell—or the infrastructure to support one—ever been found in a tower?
- If, as suggested, the towers were built in the sixth/seventh century, is that not three hundred years before bronze bells were cast?
- How is it possible for a handbell chime to be heard from nearly 100 feet up?
- How do you fit a bronze bell through the narrow doorway and lift it up to the top?

As well, if it were really the case that the Roman church was the source of the construction of the Gothic cathedrals or the round towers of Ireland, there would or should be documentary evidence in the Vatican archives describing them and where the church found the money to embark on the round tower building project, let alone the cathedral building campaigns of the twelfth and thirteenth centuries. What is listed as being publicly available in the vast Vatican Apostolic Library makes no mention of any such material, and none of my inquiries produced any useful results. Of course, that site does not include texts possibly located in the Vatican archives.

Ireland is unique in that not only have so many stones been left in place—despite the horrendous destruction of the last century and a half—but there is a tradition so firmly attached to the land and the people on it that it creates a unique backdrop on which new arrivals—whether ideas or peoples—are absorbed by the tolerant atmosphere prevalent in the land. I dare say this is common to other ancient lands (I think specifically of India) where a tradition has been continuously maintained over the ages, despite the cruelty of invaders, colonizers, and imposed regimes, but the low density of the population in Ireland and the huge number of ancient sites still accessible and apparent are rare to find.

There is also another traditional factor at play in Ireland: respect for the belief—no, knowledge—of an afterlife, and therefore ancestral reverence. If you plan on staying alive and healthy, you would not cut down certain trees, nor move, let alone remove certain stones. There is a very strong affinity with the natural world, particularly trees, stone, and water, which the invaders never managed to obliterate.

Quite when and under what impulse the towers came into being is pure conjecture, although it is generally accepted that the arrival of lime mortar made building significantly easier since it rendered reliance on a dry-stone structure obsolete, although the dating of that is uncertain. Even if the provenance and date of the arrival of lime mortar in Ireland were known, the fact still remains that the church never used it until the tenth century.

The round towers in Ireland have been in turn a subject of active discussion, a bone of contention, and a quaint but virulent Victorian dispute before they fell into the very large pot of archaeological question marks. What is intriguing is the persistence of the interest they infuse over time, with a book on the subject periodically appearing on the shelves—even if the book has nothing new to add. It is as if there is an urge to stress the official dogma that it was the church that built the towers, for that is the inevitable line spun, with absolutely no evidence to back up that supposition.

My aim is to add to that lengthening list of books but with the addition of an original theory, which can not only be demonstrated, but actively experimented with repeatedly, so it's more than just a theory, and I will be discussing it at length in the upcoming chapters.

2

Breaking Down
the Christian
Origin Theory

A s it is generally the victors who write history, one cannot help but wonder what happens when the vanquished never committed anything to writing in the first place. The story as told does not ring true in regard to events in Ireland, but in Gaul and Britain, the Druids clearly had to be eradicated so Christianity could be progressively imposed on the national psyche. We shall never know by what means the new religion placated the warring tribes of Ulster, Munster, Leinster, Connacht, and tentatively Meath, but brotherly love would seem to be a serious outsider.

There are no surviving records prior to the Annals of Ulster, which started in the year 431 CE, and as the name suggests, they only dealt with the top northeastern corner of the country. Some Roman and early Christian writers speak of the Druids, mostly in a disparaging tone, making it quasi-impossible to state accurately what their beliefs and practices amounted to, but we do have a substantial mythology and, what is even more conclusive and indicative of the mindset of those "lost" people, who committed their lore to living memory, is the legacy left in the stones.

They must have been fine mathematicians and astronomers, given their ability to erect megaliths and align underground passageways to determine key periods of time, essential for a society living in harmony with nature. Newgrange is perhaps the most famous site for its solstice alignments. There are, however, many more sites about which we are still unsure of their original purpose. While it is very manifest that these sites possess hard-to-explain facets, there can be little doubt that they were for some practical benefit.

The Druids were renowned for their wisdom and knowledge of nature, music, justice, and oral tradition. In the mind of an invader, and the Romans proved their point in Britain and earlier in Gaul, such philosopher-scientists had to be removed if a new order was to be installed.

Ireland was never invaded by the Romans for reasons unspecified. One thing is for sure, though, the latter did not like travelling, let alone fighting, on the sea, and would avoid it when possible. Most of the Celtic tribes, on the contrary, felt quite at home on the water. Prior to the Roman invasion of Britain, Julius Caesar had to resort to subterfuge when he destroyed the Celtic navy by burning their ships while still in harbor somewhere in Gaul, and he writes about it in *De Bello Gallico.*

Surely the most controversial aspect of ancient history lies in the dating of events. It would be way beyond the scope of this chapter, let alone my ability, to attempt some form of reconciliation. The mainstream version of Irish affairs of the last two hundred years have been radically perverted to portray a vastly different story from what is now coming to light; I refer, of course, to the so-called famine and events around 1916. While it is now practically impossible to make any chronological sense of the arrival of Christianity in the western isles, given the chasm of time lapsed, it is intriguing to learn of the supposed events that were key to subsequent actions and decisions, and their consequences.

Having said that, we are told that the Druids were wiped out in Gaul and Britain by the Romans. Who liquidated them in Ireland is unknown, but there is no doubt as to what filled the vacuum. The fingerprints of the materialistic Roman mindset are everywhere: absolute authority, male chauvinism (the Druids apparently welcomed both sexes), relentless expansionism, mystical ritual (to capture simple hearts and minds), and so on.

We can safely infer that the gradual implementation of these constructs would have fitted prevalent conditions, in that ruling Irish dynasties would find a much-needed support in those changing times from an apparently benign foreign source. And so, the way was opened to the Christian missionaries, who by their sincerity and simplicity—characteristics much valued by the Druids according to what the modern tradition would have us believe—gradually (a firm foothold was established by the fifth century) paved the way to the full control of Rome at the Synod of Whitby in 664 CE, which was when the variants of the Irish church were removed and the die was cast. The English contingent of the Roman tradition then, by force of numbers, imposed on the Celtic tradition of Ireland and the rest is history!

As always when looking back into the distant past, we must—if possible—be wary of the ideas we hold and try to put affairs into the context of that particular time. We use the term Druid as if it was a hard and fast social order, with rules, standards, structure, and so on. That was probably not so. Like the society of the day, it was in all likelihood a fragmented body—while no doubt sharing a great many fundamental skillsets in divination, medicine, natural science, and psychology, as well as their teaching but having to operate in the rough-and-tumble society of the day, where changing allegiances to neighboring families could make or break a bard or ovate.

Mainstream history would have it that it was Constantine the Great (272–337 CE, and emperor for 31 years) who put Roman Christianity on the map. He issued the Edict of Milan in the year 313, which made Christianity legal for the first time in the empire, and called the Council of Nicaea in 325, which codified Christian belief amid numerous sects with conflicting interpretations. Incidentally, he was baptised on his deathbed, so it is quite reasonable to question his political motivation. Quite what happened is vague at best, but a much overlooked fact is that the entire face of Christianity changed from then on. It became the state religion, and dissent was not permitted. It became a uniform creed with no place for local differences; Christians started serving in the army, thereby condoning violence in the "new" faith (the means justify the end).

There was an intense influx of Roman Christian missionaries into Ireland, and the Celtic variety with its sympathy for nature and tolerance for the old ways was inexorably removed, and by 451 CE, Patrick was top of the pile in Ireland. No mention has ever been forthcoming as to what might have happened to the Druids, their knowledge, and practices. However, as it was furthest from the watchful eye of Rome, Ireland's Christianity developed individual traditions and—to varying degrees of accuracy—has been considered idiosyncratically distinct from the Roman Catholic Church.

With this distinction made, here is a brief overview of some of the literature regarding the background on the far-fetched claim that the Catholic church was the originator of the towers, which has been repeated so incessantly over the past two hundred years that it is now fixed in the collective mind.

I have selected the following four texts of relevance to the round towers published between 1833 and 2001 to present and/or discuss here not for what they reveal, which is negligible, but to demonstrate the indoctrination portrayed: George Petrie in 1833 (but published in 1845), George Lennox Barrow in 1979, Brian Lalor in 1999, and Alanna Moore in 2001. These four, and many other texts from celebrated academics and antiquarians, persist in setting forth the establishment position of the unproven Christian origin. Within these texts, there is overt wishful thinking, a resounding absence of convincing evidence, and denigration of those with other views as seen in Petrie's opening paragraph on the following page.

GEORGE PETRIE (1790–1866)

In 1832, at the instigation of Irish archaeologist George Petrie himself, recently elevated to be a member of the council of the Royal Irish Academy (RIA), it was resolved to hold an essay competition on the origin and uses of the round towers of Ireland. In 1833 the prize of a gold medal and fifty pounds was awarded to Petrie for a twenty-page document on the topic. It is perplexing why neither Petrie's 20-page document nor his 558-page book are available on the RIA website, but his watercolors are.

Here is what George Petrie had to say in his 558-page book titled *The Ecclesiastical Architecture of Ireland Anterior to the Anglo-Norman Invasion: Comprising an Essay on the Origin and Uses of the Round Towers of Ireland* (there was promise of a second volume but it never appeared despite the fact that the author survived another thirty years):

> The question of the origin and uses of the round towers of Ireland has so frequently occupied the attention of distinguished modern antiquaries, without any decisive result, that it is now generally considered beyond the reach of conclusive investigation; and any further attempt to remove the mystery connected with it may, perhaps, be looked upon as hopeless and presumptuous. If, however, it be considered that most of those inquirers, however distinguished for general ability or learning, have been but imperfectly qualified for this undertaking, from the want of the peculiar attainments which the subject required—inasmuch as they possessed but little accurate skill in the science (if it may be so called) of architectural antiquities, but slight knowledge of our ancient annals and ecclesiastical records, and, above all, no extensive acquaintance with the architectural peculiarities observable in the Towers, and other ancient Irish buildings—it will not appear extraordinary that they should have failed in arriving at satisfactory conclusions, while, at the same time, the truth might

be within the reach of discovery by a better directed course of inquiry and more diligent research.

Hitherto, indeed, we have had little on the subject but speculation, and that not unfrequently of a visionary kind, and growing out of a mistaken and unphilosophical zeal in support of the claims of our country to an early civilization; and even the truth—which most certainly has been partially seen by the more sober-minded investigators—having been advocated only hypothetically, has failed to be established, from the absence of that evidence which facts alone could supply.

Such at least appears to have been the conclusion at which the Royal Irish Academy arrived, when, in offering a valuable premium for any essay that would decide this long-disputed question, they prescribed, as one of the conditions, that the monuments to be treated of should be carefully examined, and their characteristic details described and delineated.

In the following inquiry, therefore, I have strictly adhered to the condition thus prescribed by the Academy. The Towers have been all subjected to a careful examination, and their peculiarities accurately noticed; while our ancient records, and every other probable source of information, have been searched for such facts or notices as might contribute to throw light upon their history. I have even gone further: I have examined, for the purpose of comparison with the Towers, not only all the vestiges of early Christian architecture remaining in Ireland, but also those of monuments of known or probable Pagan origin. The results, I trust, will be found satisfactory, and will suffice to establish, beyond all reasonable doubt, the following conclusions:

1. That the Towers are of Christian and ecclesiastical origin, and were erected at various periods between the fifth and thirteenth centuries.

2. That they were designed to answer, at least, a twofold use, namely, to serve as belfries, and as keeps, or places of strength, in which the

sacred utensils, books, relics, and other valuables were deposited, and into which the ecclesiastics, to whom they belonged, could retire for security in cases of sudden predatory attack.

3. That they were probably also used, when occasion required, as beacons, and watch-towers.

These conclusions, which have been already advocated *separately* by many distinguished antiquaries—among whom are Molyneux, Ledwich, Pinkerton, Sir Walter Scott, Montmorency, Brewer, and Otway—will be proved by the following evidences:

For the first conclusion, namely, that the Towers are of Christian origin:

1. The Towers are never found unconnected with ancient ecclesiastical foundations.
2. The architectural styles exhibit no features or peculiarities not equally found in the original churches with which they are locally connected, when such remain.
3. On several of them Christian emblems are observable, and others display in the details a style of architecture universally acknowledged to be of Christian origin.
4. They possess, invariably, architectural features not found in any buildings in Ireland ascertained to be of Pagan times.

For the second conclusion, namely, that they were intended to serve the double purpose of belfries, and keeps, or castles, for the uses already specified:

1. Their architectural construction, as will appear, eminently favours this conclusion.
2. A variety of passages, extracted from our annals and other authentic documents, will prove that they were constantly applied to both these purposes.

For the third conclusion, namely, that they may have also been occasionally used as beacons, and watch-towers:

1. There are some historical evidences which render such a hypothesis extremely probable.

2. The necessity which must have existed in early Christian times for such beacons, and watch-towers, and the perfect fitness of the Round Towers to answer such purposes, will strongly support this conclusion.

These conclusions—or, at least, such of them as presume the Towers to have had a Christian origin, and to have served the purpose of a belfry—will be further corroborated by the uniform and concurrent tradition of the country, and, above all, by authentic evidences, which shall be adduced, relative to the erection of several of the Towers, with the names and eras of their founders.

Previously, however, to entering on this investigation, it will be conformable with custom, and probably expected, that I should take a summary review of the various theories of received authority from which I find myself compelled to dissent, and of the evidences and arguments by which it has been attempted to support them. If each of these theories had not its class of adherents I would gladly avoid trespassing on the reader's time by such a formal examination; for the theory which I have proposed must destroy the value of all those from which it substantially differs, or be itself unsatisfactory. I shall endeavour, however, to be as concise as possible, noticing only those evidences, or arguments, that seem worthy of serious consideration, from the respectability of their advocates and the importance which has been attached to them.

These theories, which have had reference both to the origin and uses of the Towers, have been as follows:

FIRST, as respects their origin:

1. That they were erected by the Danes.

2. That they were of Phoenician origin.

Secondly, as respects their uses:

1. That they were fire-temples.
2. That they were used as places from which to proclaim the Druidical festivals.
3. That they were gnomons, or astronomical observatories.
4. That they were phallic emblems, or Buddhist temples.
5. That they were anchorite towers, or stylite columns.
6. That they were penitential prisons.
7. That they were belfries.
8. That they were keeps, or monastic castles.
9. That they were beacons and watch-towers.

It will be observed, that I dissent from the last three theories, only as far as regards the appropriation of the Towers exclusively to any one of the purposes thus assigned to them.

Twelve years after winning the first prize for his meager twenty-page contribution, Petrie published a book where from pages 5 to 122, he refutes the theories mentioned above, and then from page 123 onward expands on the features of Christian architecture and decoration in Ireland and elsewhere, with *absolutely no application* to the round towers, let alone any mention of their origin and use. Only as of page 358 does he expound on belfries when he offers more obscure arguments in support of his preferred theory. It could be usefully recalled at this point that a belfry was designed to house a bell, generally made of iron coated with copper. Although the Romans apparently used bells in London to announce the time, the belfries were probably in open but populated areas, which does not fit in with the locations of many, if not most of the towers in Ireland. Built in the late eleventh century in France and Flanders especially, bells were destined for urban areas where local communities could be reminded of church services, time of day, emergencies, or summons. The larger bells often required several men to ring them, they were so heavy. Few and far between are the round towers located in urban areas, especially of that period, and even fewer with a doorway large enough to allow the passage of a heavy bell that would then have to be hoisted up the remaining sixty-odd feet.

One can legitimately wonder why for more than a hundred pages Petrie deals with the designs on capitals and bases of columns, as they are totally absent from a round tower, but he insists on inundating us with his architectural expertise, while regrettably adding precisely nothing to the matter in hand.

Given his constant reference to the findings, measurements, and deductions of other researchers, it appears that Petrie relies on the observations of the same, and he perhaps spent little time inside a tower, especially in view of so many anomalies in his book:

1. He never mentions the double wall construction employed in all authentic towers.

2. He maintains that the protuberant base of the Clondalkin tower is original. To support this, he compares it to a tower with a similar base as portrayed on a a seventeenth-century seal. However that tower was in distant Wales and apparently blown down in the thirteenth century.

3. He claims the towers were topped by a stone cross, but one has yet to be found.

4. He states that as no pagan building in Ireland resembling a round tower has ever been found that only Christian builders had the necessary skill.

5. He further states that prior to the arrival of Christianity the Irish did not have the know-how to make lime mortar, nor to make an arch despite "innumerable" remains of buildings. He omits to list the remains and in the absence of proof of such, it makes for a shaky foundation to lay such a broad claim.

6. Although he frequently admits that the ecclesiastic building(s)

in the vicinity of the tower were built several centuries after the tower, he never addresses the why or wherefore.

GEORGE LENNOX BARROW (1921–1989)

Apart from the intense work Barrow did over a period of eight years or so on his book *The Round Towers of Ireland*, he added nothing new to the paradigm—in fact he persisted with the same story that the church was at the origin of the towers despite stating, "Documentary sources on the towers are few and unreliable." He built a very useful gazetteer or compendium of towers actually in existence, which had reputedly existed or were referred to in the annals, with photos, drawings, measurements, and local lore. His references are uniquely drawn from sources that uphold the mainstream paradigm, ignoring or ridiculing any texts that challenge that. For example, he dismisses peremptorily the work of Henry O'Brien, with "this farrago of romantic and mystical nonsense." Not exactly the type of open mind one might reasonably hope to find in the twentieth century with regard to the only other person to take part in the RIA competition of 1833, winning second place, £20, and a bronze medal for his 509-page document. Especially as O'Brien was only twenty-five years old giving his all against a man of fifty-three who managed a rather paltry twenty pages.

In this book, there is something very disturbing about the entry for Kilmacduagh, which was restored in 1878–1879 by the Board of Works. The entire complex of ecclesiastical buildings was cleared and revamped, as was the graveyard, and work was done on the tower, supposedly to install a lightning conductor. No mention is made of the state of the tower by Barrow; however he cites extensively from an undated letter to the editor of the *Irish Builder*, from an anonymous J. A. F., where one learns on page 29 that the tower was fast decaying like the other buildings and that the tower was completely restored, as "a great portion of the south side had fallen; and there were evidences to justify the gravest apprehensions regarding a considerable portion of what remained." It is explained in the letter that the interior of the tower was excavated in order to install the conductor, which led to the discovery of human remains (four bodies in all, apparently) below the level of the foundations. The archives of the *Irish Builder* can be found on the Internet Archive website.

Now while we don't know how far down, or to what elevation of the tower, the barrel was reduced, an interesting clue can be found in some of the documentation from Ireland's earliest archaeologists. Lord Edwin Richard Whydham-Quin, third Earl of Dunraven (1812–1871) was an archaeologist who employed early photography in addition to illustrations in his investigations. Dunraven's photo of the Kilmacduagh tower shows the east-facing window on level five to be intact. From there to the earthen floor is a distance of almost 53 feet, and from the earth floor to the soft soil where the skeletons were found is 19 feet. So almost 20 feet of earth had to be dug and raised 25 feet to remove it from the inside of the base, with the last 6 feet or so of depth being dug out from a roughly circular hole in the stone foundations that measured 5 feet on average. Clearly not large enough for anybody but the smallest of people, and four of them at that, with some of the skeletons protruding out beyond the stone foundations.

Predicting the time for a set of bones to revert to the earth is not a precise science as it depends on the acidity of the soil, humidity, the condition of the body, and so on. However, if this tower was built in the ninth century—two hundred years after the earliest building date estimates—that means it took a thousand-odd years for the soggy, acidic soil to *not* break down the skeletons of four pagans. A man of the cloth would not have gone digging around in a graveyard to build a tower is Petrie's argument. Either the pagans were made of sterner stuff than us, or there is another option, such as a rapid disposal of genocide victims from the 1850s era at the foot of a tower in a graveyard.

....

Ei incumbit probatio qui dicit, non qui negat.
The burden of proof lies on he who affirms, not on he who denies.

....

BRIAN LALOR (1941–)

The second in the dogmatic, must-be-church series is *Ireland's Round Towers: Origins and Architecture Explored*, written by Brian Lalor in 1999. On the very first page, one reads this opening statement: "Round Towers are the only form of architecture unique to Ireland. . . . An informative guide to these intriguing products of the Celtic imagination."

It makes it hard to have any confidence in what Mr. Lalor might have to say thereafter if he is not aware of the towers in Scotland, which he obviously is because on page 96 of the same book he devotes half a page to the existence of the Scottish towers and the one on the Isle of Man but offers this strange statement: "None of these buildings can seriously be considered as significant influences on Irish early medieval tower building. None retains a conical cap, and as all appear late in terms of the door typology, cannot be considered as forerunners to the *cloicteach*, but as parallel developments or evidence of the influence of Irish building practice."

"The only two towers with complete caps . . . are Clondalkin and Devenish, and only the former *appears* to be in its original state." (Emphasis is mine.)

Because the tower does not have a conical cap, it is dismissed despite the fact that it is a well-established fact that few if any of the towers have their original caps! They have been reworked, perhaps restored in some rare cases, but definitely modified in the past few centuries. We have absolutely no idea what the original form might have been.

The "products of the Celtic imagination" would imply that Mr. Lalor knows something that no one else is privy to. Generally, common sense and as of recent times, when an out-of-place object is found anywhere, the scientific approach is to search, to cast the net far and wide so as to bring as much light as possible to the enigma. A shame. Incidentally, if anything is out of place in this context, it is the ecclesiastical buildings in proximity, sometimes even attached. Was the rest of humanity outside of the Celts deprived of imagination? It does seem like one more effort to push the Celts as being somehow at the origin, which is almost as absurd as to suggest it was the church.

When one reads this type of prose, one comes away with the very strong impression that new ideas are clearly frowned upon, even ridiculed. That is a very manifest sign that something needs to be concealed, the familiar "Nothing to be seen here, move on!"

Neither Lalor nor Barrow ever mention the Pelasgian or Cyclopean architecture that is often employed in the towers. This technique is found across the ancient world and can be seen at sites such as Giza, the Acropolis of Athens, and Machu Picchu. It was Henry O'Neill in his brief but fact-filled fifty-six-page document *The Round Towers of Ireland*, dating from 1877, who brings this to our attention, especially as concerns the Lusk round tower.

ALANNA MOORE

Another author to have written a book on the round towers is Alanna Moore. Despite her useful input on the practical aspects of model towers made of sandpaper, she has a tendency to fawn over the established opinion, confounding her belief with what the authors above have to say. Often even adding rather obscure statements, such as "there would have been a monastery neighboring those towers found in isolation," without providing any support for the statement, and "*Danu* is the word in Sanskrit for streams of water." I'm not sure where that one came from but definitely not from *Monier-Williams Sanskrit Dictionary*.

Moore claims that there are seventy-five towers in Ireland despite the fact that she frequently refers to Brian Lalor's work which lists only seventy-three, including vanished towers! She finds Barrow historically inaccurate because he dated the towers to pre-Christian times. And even stranger, she states that George Petrie precedes Henry O'Brien despite the fact that Petrie, as a member of the RIA, first called for papers on the origin and purpose of the round towers in 1833, to which O'Brien responded. She does not come across as very factually convincing when she further states her belief that the Atlanteans may have been an advanced amphibious race.

Well okay, but quite how that fits in with the subject supposedly at hand, I fail to grasp, and even less how she can claim that Brian Lalor wrote "the most comprehensive tome on round towers to date" given these omissions.

What's more, so much New Age myth is asserted in her book *Stone Age Farming*, it makes for a very confusing swamp of information, techniques, and fable that cloud an already murky horizon. Moore is considered to be a master dowser, and since the beauty of dowsing is that it makes it possible to acquire definite yes or no answers to precise questions, one might reasonably expect her to make a clear statement as to what might be below the ground of each and every round tower. However, that is not to be and one is met with a hash of information impossible to confirm or invalidate, serving to do little save generating further questions and stimulating the sensationalism.

ADDITIONAL RESOURCES

There are a number of books that would be worth reading to obtain a less-biased view of the towers than the ones mentioned previously. It might be helpful to start with the rector of Ennis, Marcus Keane, who wrote a most useful book titled *The Towers and Temples of Ancient Ireland: Their Origin and History Discussed from a New Point of View* in 1867. For a man of the cloth, Keane is way out there, investigating and discussing with numerous literary references and pictures in support of a whole range of ideas in the aim of discovering, not imposing.

The aforementioned *Round Towers of Ireland* part 1, written by Henry O'Neill in 1877, is a very brief (fifty-six pages) but useful study on the towers in the Dublin area, including Clondalkin, Lusk, and Swords. Unfortunately, part 2 never saw the light of day.

The Round Towers of Ireland: Or the History of the Tuath-De-Danaans was written by Henry O'Brien in 1898. This is the author mentioned earlier who produced an extraordinarily rich document with a very varied reference base, proposing that the towers were fertility or phallic symbols (called *lingam* in Sanskrit) employed in an ancient form of the veneration of fire or sun and moon worship introduced by Asian Indians.

The Mysterious Round Towers of Ireland, written by Thomas Sheridan in 2018, is a very eloquent plea to take a new approach to the towers and to move away from the prevailing assumption and theory upheld by academia and the church. His arguments, particularly the question as to why the towers were not destroyed by the early Christians as pagan symbols, are convincing and forceful.

THE ROMAN CHURCH ESTABLISHMENT

What can one expect from the church establishment? The very word speaks volumes on the epithet. Something established does not move. We humans love such organizations, they make us feel secure and enable us to feel that everything is in its place. Somewhat like God's garden—an appealing figment of the imagination depending on order, structure, discipline, and most importantly, belief.

The belief that such an establishment structure is the most suitable solution to organize the material world is a very common one, and why not. There is probably no alternative in actual fact, however free-thinking we may like to believe ourselves to be!

Problems arise when the hierarchy is upset, for example when a closely-held belief is questioned. This would apply to all domains of the establishment—medicine, academia, religion, the press, and publishing—for there is no separation among the component elements; they are mutually dependent in their need to maintain their hold over society. This is merely an observation, not a judgment, for there is absolutely nothing I can do about it, even if I felt so inclined.

The truth, however, has a habit of coming to the surface, even if no one is looking when it gets there. But if there is a deliberate effort to make it sink again, to remove it from public view by obfuscation, ridicule, or bulldozing, we can reasonably assume—due to past experience—that there is something authentic going on and our attention should be piqued.

Without going to the extreme of saying that the establishment has made every effort to conceal, denigrate, and deride Irish culture in general, it would be quite fair to say that certain aspects of Irish history have been eclipsed in order to diminish the contribution played by the inhabitants of that island—irrespective of their tribal appurtenance, creed, or politics.

The church (the Roman version) took a long time to work its way to the top of the establishment pile and as is systematically the case when a group of men gather (especially when the women have been evicted), their egos encountered no restraint in achieving their short-sighted aims, and they were obligated to step on their original human values, crushing all resistance by whatever means at their disposal. If one can piece together the fabric of the Irish form of Christianity, scattered as it was and no match for the Roman (church) bulldozer, much like their legions of a former epoch, it appears to be a more humane version. The religious schools of Ireland were all the rage in western Europe for several centuries, with students coming from far and near to learn, and their alumni traveling throughout Europe becoming saints and renowned teachers.

One thing is sure, and even the establishment experts agree, that despite strong evidence that the towers have an Irish origin, there is *no* documentary evidence to support the theory that any church, let alone the *Roman church,* was the originator of the round towers. So, if we could start anew and look at what we still have, and contrary to what has been written to date, look at the *why* rather than the when or by whom, we stand a better chance of learning what the practical applications of these remarkable structures were, and far more importantly, *are.*

3
The Use of Stone
by Humans

One does not need to be a devout Catholic to be stirred by the majesty of a Gothic cathedral. The same can be said for being in the presence of a megalith, especially an array of stones such as Carnac in Brittany, Stonehenge in the UK, or Loughcrew in Ireland. These are awe-inspiring monuments.

But what do we truly know about stone itself? What is it that resonates within us when we are in the presence of a stone circle? Is there something there that we sense that somehow enables a connection? Is it simply our complete lack of education about what that sense is that causes us to relegate the experience to one more overlooked gut feeling?

Despite the extensive use of stone in just about every culture throughout the world, there is little or no information available to help us understand its properties, let alone how those characteristics can be, or have been, applied. This book started as an essay, an attempt to clarify what I had been researching and experimenting with for a number of years. A chance discovery while examining one of the Irish round towers then led to a winding but essential journey along the majestic paths of Pelasgian or Cyclopean architecture, civilizations long past with their occulted yet intimated techniques, as well as a few dead-ends, which had to be explored so as to exorcise certain possibilities. All due to an apparently banal and simple insight regarding stone: it is magnetic.

While the thrust of what I am proposing here is multipronged and will hopefully be dealt with in detail in due course, most of what I say is based on simple observation, historical and scientific research, and experimentation, with the addition of an inevitable element of intuition as we are dealing with (publicly) undocumented, obscure (perhaps because they are obscured) topics, such as the following:

- There is a magnetic field of force present in stone.
- Stone can be used to generate a feel-good sensation.
- The varying height of the towers could very well act in the same manner as antennae, which have a specific length in relation to their function.
- Geographic positioning of stones over geological phenomena has differing effects.
- Pattern or layout enables resonance.
- The interaction between cosmic and telluric forces can be focalized.
- Energy can be beneficially modified by using the characteristics of specific locations.
- Energy can be beneficially modified by using specific materials.
- Shape and pattern can and does alter the flow of energy's movement.
- Stone can provide many benefits to health—ours and that of the environment, and there are specific means to achieve those benefits.

For some traditions, especially in China and India, stone is considered the origin of water since it comes from the depths of the earth. In the 1960s, Stephan Riess (1898–1985), a German geochemist, proposed that water is produced deep inside the earth in crystal rock, with his primary water theory that is progressively gaining traction. Ever since Jacques Benveniste published his controversial article in *Nature* magazine in 1988, in which he suggested water has memory, a trend has been set. He didn't go as far as Walter Schauberger, who claimed it also has its own intelligence, but it was still radical enough for him to be ridiculed and lose his job. He was vindicated by Luc Montagnier some years later when the latter made the same statement, to very much the same reaction from the scientific community. No doubt, in a few more years, research and experimentation will add some clarity. We are apparently doomed to remain in ignorance until such time that we take a more holistic view of the intimate relationship between stone, water, and the movement created by magnetism, for there is some kind of partnership at work with this trio—a threesome that is deemed beyond scientific concern being so insignificant and of such little promise. One can well muse why water is the supreme symbol of humility, always taking the lowest position. There is basically zero research into low-field magnetism, yet the "energy" from a hand can relieve chronic pain, as many a magnetiser or Reiki practitioner knows. As for stone, it is the very foundation of our home—Earth. And who does not feel a sensation of insignificance in the presence of a mountain looming above you? Behold the dimensions of mystery in these banal objects of water and stone!

Prior to 1949, Japanese and Chinese landscape gardeners used the properties of magnetism—namely, diamagnetism and paramagnetism (more on these in the following chapter)—to generate an enhanced environment for plant growth, but not just for the benefit of the plants. That sense of well-being was also passed on to humans and animals because they, too, are connected to or share the same vital force we all benefit from. I would propose that there is a very strong chance the ancients were deliberately doing the same when they constructed the monuments of stone that we still have among us today.

Unfortunately, or so it seems, there has been a constant and intense effort throughout the ages to occult the importance of the magnetic field for one and all of the earth's flora and fauna. Or, as Abraham Liboff said in his 2013 article in *Electromagnetic Biology and Medicine,* "Why are living things sensitive to weak magnetic fields?" And then further on, "The interaction of weak magnetic fields, with intensities on the order of the geomagnetic field, is a very interesting subject that only recently, in the last few decades, has received much scientific attention."

I would put my money on ignorance and neglect. It is just too obvious that stone has magnetic properties, but if you do not know how to detect and then deploy them, they fall into disuse and the trash can of history. Without instruments to measure the low magnetic fields of force, it is nigh on impossible to prove to the satisfaction of one and all that a field of force impacts the individual's nervous system, even though such a sensation might well be felt physically and the beneficial effects experienced. Such measuring devices are not commonly available apart from some very expensive laboratory equipment, which is not readily portable. Having said that, it is a well-known phenomenon among alternative therapists that magnetism as well as the laying on of hands of someone with a strong magnetic field induces not only relaxation and a sense of peace and calm but also sometimes a cure.

Under any regime, at any period of history, that sense of peace is a much-prized quality valued by any society with a focus on harmony and well-being, even if we do not understand how such resultant well-being is accomplished. It is important to note that this sense is not just a faculty specific to people; it is also commonly found to be a much appreciated facet of any site, building, or location that generates it.

In most cultures such locations become places of pilgrimage, power spots, or sacred sites, and their domination can become a way of controlling of the surrounding population, or at least their attitude toward the source of that influence. There are numerous examples of this found around the world, especially with water sources that have healing or therapeutic properties. As we shall see further on, I think this is very much what happened when the church claimed authorship of the round towers, but as mentioned earlier—the burden of proof lies on he who affirms, not on he who denies. The only difference being that it is the supporters of the church that claim authorship whereas I merely request the proof—to no avail.

You might agree that one of the most impressive effects on entering a cathedral built during the medieval period, from 1170–1340 or so, is a sensation that is almost physical in its intensity. That sensation is not found on entering any other ecclesiastic building, or in the same manner, whereas it *is* encountered when crossing the threshold of certain ancient monuments that share some of the principles discussed here. Once again we are very much in the dark as to a reason, if any, for this.

As Philip Ball points out in his book *Universe of Stone*, the Gothic cathedral for Victor Hugo "was a social construction, a temple made for and by the people rather than decreed by an ecclesiastical elite." There are few pilgrimage sites in western Europe that are not dominated, controlled, or claimed by the church, yet we never question that authority. While it is possible that could be the case, would it be too much to ask to have an explanation, if not some evidence?

Literature is indeed most practical—when we know what we are reading—but time is not tender on papyrus, paper, and leaves, so limits are soon reached, and we must revert to the stones to gain insight. When we are fortunate enough to have them available, and they have not been removed, destroyed, buried, or used for other purposes, we have an open book before us. But what is the message? Can we even read the script?

It is largely thanks to archaeological remains that we have a vague idea as to what our forebears were up to, as there seems to be little in the way of explanation from literature or mythology in any of the European traditions. In actual fact, there is nothing more than conjecture for saying that, and it is thanks to people like Alexander Thom (1894–1985), the Scottish researcher who not only did so much to demonstrate that the ancients were highly competent engineers and technicians but also made a space for archaeoastronomy, so proving that the standing stones serve a very distinct purpose, which regrettably is not accepted as gospel truth by mainstream science. As a result, any ideas coming out of alternative archaeology can only be considered hypotheses, which is what they have always been, and what is written here is no different save that it is confirmed and actively used in the Chinese and Indian traditions of feng shui and vastu. Accident, coincidence, synchronicity, and design are some of the expressions we employ to explain what we do not—and may never—understand in life. These pages are no exception, as they recount a story of revelation and discovery of an unusual property of stone.

This hopefully makes for a more convincing argument backed as it is by practical applications, which can be easily experimented with, verified, and applied and hopefully contributes a dimension that is rarely if ever considered and, as a consequence, is simply rejected. There is no end to the skepticism born of modern "science," and regrettably, no space granted to the sincere holistic approach requiring an open mind, which at the worst could provide a little insight.

It would seem that in ancient times, the potential of some of these principles was better understood than it is today, and practical use was made of them. It goes without saying that humans have a streak of ingenuity, and from our vantage point of the present moment, we look back at various stages in the past where people have used that inventiveness to incorporate the forces of Nature into their structures and practices in much the same way that is done today—well, more or less. The difference being that we now know that interfering with planetary frequencies is fraught with consequence. There will always be those who, with the benefit of hindsight or due to an innate common sense and decency, keep within the bounds of reason and maintain a healthy respect for Nature. By the same token, there will also be those determined to profit at any cost.

I can only echo Alexander Thom as he said in *Megalithic Lunar Observatories*, "We do not know the extent of Megalithic man's knowledge of geometry and astronomy. Perhaps, we never shall. He was a competent engineer. Witness how he could set out large projects to an accuracy approaching 1 in 1000, and how he could transport and erect blocks of stone weighing up to 50 tons."

The focus of this book is on the "lower end" of the monumental scale, which includes the vast majority of stone structures, starting with the more common standing stones, obelisks, circles, menhirs, and round towers rather than the gigantic structures such as the pyramids found in Egypt, North and South America, China, Java, Bosnia, the Bermuda Triangle, Greenland, and so on.

Archaeologists, like every other "ist" generated by modern science, are often obliged to restrict their investigations to the academically correct and acceptable criteria imposed by current dogma and frequently by tenure. Few academics dare venture outside that corral. There are some amateurs who have done and do so, but rarely do we hear of them in the mainstream, if at all. Obviously, that is a great shame because so many brilliant and curious minds are stymied from casting their insights into the public arena, and the fictional narrative is maintained. As in modern-day society, there is no longer room for an arena where open discussion can be held of ideas and discoveries that might be of benefit for the world at large.

In much the same way that no sensible Egyptologist—wishing to retain their livelihood—would mention that no bodies have ever been found in a pyramid, no orthodox researcher into the Irish round towers would ever dream of not mentioning a church in connection with the towers.

As a holistic researcher, specializing solely in extracting a version of the truth from the evidence, using means both orthodox and unorthodox, I am free to roam at will outside the box. Any challenge to what is written here is most welcome so long as it is constructive (nonderogatory), enabling me and others to better understand, and conducive to greater precision in the subject matter.

At the end of the day, the most apparent feature of ancient human history is to be found via the use of stone. Throughout the known world, stone vestiges are to be found and reveal all sorts of information as to what life might have been like years ago. However, there is far more left concealed from our understanding, and if we persist in neglecting the factors, some of which are discussed here, we are doomed to remain in the dark.

4

Magnetism

Its Function and Inherent Properties

A pragmatic definition of *magnetism* would be useful to guide us, but there is none to be found. It is as if the subject is booted into touch, to be dealt with at some later stage, another day. The orthodox scientific reference Encyclopedia Britannica attempts with the following remarkably lame explanation: "All matter exhibits magnetic properties to some degree." Presumably this refers to the properties of attraction and repulsion. There is, however, no explanation of where it came from or how it got there in the first place. Instead, we are fobbed off with suppositions, theories, and mind-boggling data, whereby monopoles, dipoles, electrons, and atoms torque and spin through a bewildering array of the human-denominated categories of physics, biology, chemistry, geology, anatomy, and other branches of "science," plus others for which we have not yet found names, all carefully measured in tesla or gauss. Welcome to the world of quantity!

In other words, as is so often the case in the current era, instead of assuming a potentially holistic, open-minded approach with a broad consideration of an interrelationship with life in general, we isolate and deal with a single function, so that it can more readily be expressed as a quantity, but in so doing, we basically kill the event, in much the same way we do when we place a cell under a microscope and study it outside of its natural environment.

Of course, having discovered a facet of a phenomenon, it is an excellent idea to research and develop it within the limits of one's capacity. But to curtail further study is suspect, and it is hard to imagine a lamer statement than "all matter exhibits magnetic properties to some degree." One cannot help but wonder about energy-directed weaponry and the applications of electromagnetic frequencies, which rely on a significant "magnetic" component in order to function.

Once again, we are obliged to beaver away behind the scenes to find ways to restore a little harmony in our environment because it is clear that neglecting the need to maintain a "magnetic" balance has now become very damaging for Nature at large—humans, animals, plants, and the atmosphere specifically.

The word *magnetic* has been deliberately placed in quotation marks above as it is too vague an epithet for what we are trying to discover, so we need to find another term that better defines the various characteristics that make up this complex idea. We have a tendency as humans to confuse the properties of an object with its functions. In doing that, however, we lose sight of the inherent characteristics as we become absorbed, as it were, by the power of the thing in question. Especially when that thing exhibits a capacity we are interested in, to the extent that we are able to ignore the disadvantages, especially when profit is involved in the application. In this day and age, the perfect example of this would be the mobile phone.

By the same analogy, what we know as magnetism has become something so banal that we have ceased all research into the how and wherefore and fall for the lame excuse, "We don't know exactly what it is, but we know that it exists."

ORIGINS OF MAGNETISM

The thrust of this section is to propose that the movement of energy flowing in from the cosmos, its subsequent flow through the earth's strata and components, and its accumulated effect on the earth are the probable causes of this phenomenon we call magnetism. What could be more natural, from a purely functional and energetic viewpoint, than for a force to be developed when the sun and moon—the former radiating a very powerful electromagnetic force and the latter containing a yet-to-be-explained presumably magnetic capacity—have been revolving around our heads for several billions of years? When you know that an electric current can be generated by turning a magnet wrapped in a coil of wire round and round, a little perspective is gained, or should be. The earth is/has become a magnet, the source of a power concealed in its entirety because we refuse to attribute a recognized, commonplace phenomenon to our own home. The circular movement performed by the two luminaries, day after day, has not been examined in this dynamo light. There is a constant field of force evolving, unseen, unheralded but in plain sight.

Which came first, the magnetic material or the revolving force? I don't know and do not care to know. What is important in this context, I believe, is the ultra-simple fact that stone, which makes up the majority of our land mass as well as the firm bit below the seas, is affected to varying degrees by this force, as witnessed by what we know as the magnetic properties of attraction and repulsion. Of course, those two qualities are not unique to the attractor or the repeller, they are shared and interact as a function of their position to one another, with other factors most probably playing their part and affecting the intensity. We will look at this characteristic concerning the degree of strength along with the properties and uses further on.

NATURE OF MAGNETISM

There is every reason to believe that magnetism could well be the primary force at work with regard to the development, maintenance, and potential destruction of our environment. The trouble is if we do not have a minimal understanding of the sources, capacities, functions and effects of the various, apparent and invisible, energy-forming components in our environment, it is hard to apply a theory in order to determine if further experimentation is worthwhile. The constant flux of cosmic vibrations from the universe interacting with the various components of our atmosphere should not be discounted as a cause of a potential charging of the component with magnetism. The constant flow of billions of years of cosmic frequencies must have had some effect, and presumably a lasting one at that. But as with all tides, there is an ebb and flow.

According to received (rather, imposed) science, the making of a magnet, or any body showing a magnetic force, involves the aligning of the atoms in that material. This, we are told, comes about due to the spin of the atoms as they are impacted by the flow of the current to which they are sensitive. The electrons all spin in one direction, with those at one pole spinning to the right, and those at the other pole spinning to the left. This apparently depends on the number of atoms that can be magnetized.

There is no mention of an inherent form of magnetism in this formal version of science shared in academia. The method of quantitative measurement, or the gauss measurement of magnetic force, is a function of the number of molecules in the material whose electrons can be magnetized or polarized. Absolutely no attention is given to any of the other properties of magnetism beyond the measured scale; they are sublimely ignored and not given any space at all in the equation.

While it is not within my ability to explain the physics of such properties, and I do not have the infrastructure to prove their existence with experimental evidence, the fact remains for mainstream science and alternative technology that magnetism is one of the fundamental forces at work on our earth, and its patterns and fashion of functioning deserve greater attention if we hope to surprise the truth from it.

One of the main points in this book is the essential importance of magnetism. Essential because it is one of the fundamental components in the functioning of all that lives and is even present in what we consider dead. It would not be a lie to say that what we term "magnetism" is vastly underestimated for a resoundingly simple reason: its cause and prevalence in everything that exists is either neglected, ignored, or has been expressly occulted. That might appear to be a very provocative statement, I know, but when one considers that basically no research has been conducted—publicly at least—on magnetism, then there is justification to ask why, given that it is such a commonplace yet seriously misunderstood phenomenon.

That said, science has done some research on a form of magnetism in the human body which is expressed as a very low current that is a result of the flow of blood, nerve, and vital energy. Modern medicine measures this electric current in the most basic of diagnoses with the ECG or EEG. In 1992, an abstract of a paper titled *Measurement of Magnetic Field Produced from the Human Body* published in *IEEE Translation Journal on Magnetics in Japan* stated the following:

This article describes the present status of research on biomagnetism, an interdisciplinary field of research involving biology, engineering, medicine, physics, psychology and other areas. Biomagnetic fields are caused either by electric currents in conducting body tissues such as the heart, the brain and muscles, or by magnetized material in lung contamination. These magnetic fields, although measurable, are so extremely weak that a superconducting quantum interference device (SQUID) magnetometer with ultra-high sensitivity is needed to detect magnetic flux generated outside the human body. This paper mainly focuses on the remarkable progress of research on biomagnetism involving magnetoencephalograms (MEGs) and magnetocardiograms (MCGs), and on the introduction of SQUID systems for measurement of biomagnetic fields.

On December 26, 2009, a paper published by PTB, the National Metrology Institute in Berlin, in conjunction with the NIST, the

National Institute of Standards and Technology in the United States, found that an optical magnetic field sensor substantially different from the SQUID was suitable to measure biomagnetism in the picotesla range. Well, at least science now has a basis for some kind of standard.

So many aspects of this magnetic force can be quietly adopted and assimilated, providing huge advantages in our daily existence. The instances of quality, as I would call them, can be easily adopted to make life a joy rather than a bane. For instance, one can accumulate a magnetic charge by performing my magnetic recharging exercise, including tai chi or qi gung, both of which are designed for that purpose and result in reinforced vigor. Another example is the exchange of magnetism as it occurs in such practices as the laying on of hands, in which a transfer takes place, often with surprising therapeutic results. The benefits of magnetism can also be experienced through grounding oneself by walking barefoot on the earth or using a grounding mat beneath the body during sleep or a brief period of rest.

The next layer in this onion of magnetism would logically be what we call, rather obscurely, polarity.

MAGNETIC POLARITY

The central position of the North Pole in relation to its associated star, Polaris, a.k.a. the North Star, can be readily appreciated by setting a camera at a very low speed with a long-time exposure pointing at Polaris. The pole star remains stationary and central in the image with all the other stars forming circles around it. This would indicate a number of things: firstly, that the earth is not spinning at 67,000 mph or so as we are told, or the camera could not have remained focused on Polaris, and secondly, Polaris is spinning at the same speed, which would be hard, if not impossible, to explain. It further indicates that the stars and the planets rotate above our heads, thereby providing another proof of stellar and planetary movement in a circular fashion over the earth and causing the dynamo effect mentioned earlier.

The magnetic attraction as exercised by the North Pole has a definite phenomenal effect of drawing a lodestone, iron, needle, or magnetized substance. From time immemorial, that property has been used, as it is today, in all navigation systems. This is not to say that there is a so-called magnetic force exercised by the pole. To be effective, a dynamo effect must operate on a fixed axis, and that is what—I would maintain—is the line of force between the North Pole and the north star, Polaris.

This phenomenon of attracting a compass needle is only implemented from the northerly side of the direction in which the compass is held. The opposite end of the needle is not attracted to anything; it simply follows diagonally opposite the alignment, without bending or fluctuating movement as the needle would do if it were really subject to two attracting forces. The so-called South Pole, however, cannot reasonably even exist without bending the compass needle.

We are under a major illusion if we believe that the South Pole of a compass needle functions under the influence of a power of attraction coming from that area; it doesn't. The sole power of attraction in a compass is coming from the North Pole, which draws the magnetized needle toward itself.

To add to the confusion, when considering the reputed magnetic properties of the poles, we are told that in actual fact the North Pole is pointing to the south because it is *attracted* to the south, and so north is indicated by the south-seeking pole. The complexities of the above statements are immense because that would imply a need to revise the entire scientific paradigm with regard to gravity, fluids, gases, and mass, to say nothing of biology, geology, and the behavior of water.

The magnetic influences will remain a mystery so long as we do not assume a holistic approach to research and total environmental relationship. Having said that, experiments have shown for example, that exposure of earthworms to the south polarity of a magnet, rather than the north polarity are deadly; after a three-day period, the worms were dead and dehydrated. Even worse, it was found that when chickens were exposed to that polarity, they became cannibalistic and even attacked other animals, including cows! One can deduce that removing the harmonizing influence of the north pole of a magnet can lead to trouble. Consequently, the balance is best ensured by limiting excessive magnetic influences. This balance is of course severely compromised with the generalized use of electricity, radiotelephony, and modern communications.

So, irrespective of the electron spin of the one polarity being in one direction only, things do not fall apart as a result of the repulsion. On the contrary, due to their form of movement—in a spiraling vortex—they travel in both directions and stay together as a coherent whole. This can be seen in a relatively simple experiment with a slide under a microscope, a few drops of diluted sulfuric acid—or even better, whole blood—on the slide, and a magnet at either end of the slide. By switching the polarity of the magnets, the spin of the acid or blood will be seen to reverse. It can be compared to the yin and yang of Chinese metaphysics, but there is no need to create a nonexistent zone and claim that it enforces a phenomenal energy. The danger of postulating fictitious fact while ignoring the operation and form of movement of energy is great, and it is to be denounced.

Yet, apparently, we still do not understand how magnetism actually functions, and we are blithely informed that it can be explained away chemically as being an affair of paired or unpaired electrons. How sad to pass over with such nonchalance a living force that provides us with or deprives us of life!

When we consider that this magnetic force has been enclosed for billions of years in a relatively (if not completely) hermetic container, there is a strong chance that it will have created an environment that produces stuff. It has and it does—us and everything we know. Thanks to this mechanically simple but immensely complex mechanism of which we are a totally integrated component, the whole phenomenon stays on track.

I would maintain that the ancients were very aware of this force and expressed the awareness in their fashion. We know, or rather can infer, that benefits were and are derived from stones and things made of stone. For the simple reason that stone can be a highly magnetic substance, with a positive and negative polarity.

Consequently, and from now on in this book, reference will be made to the positive and negative polarity—respectively, the yang and yin characteristics—of material, stone, or human. This will avoid the confusion associated with so-called north and south poles, or polarities.

These are two totally different aspects of the phenomenon, and due to the lack of detailed research they are lumped together as if they were one subject. The power of the magnetic attraction from the North Pole is one thing. The vorticity or spiraling motion caused by the flow of movement of energy in its multiple forms is another, and most probably what creates the positive and negative, or opposing, forces, although not necessarily the power of attraction and repulsion because they accompany each other at all times and in all places and in varying degrees of strength.

That spiraling motion is found throughout Nature and is probably the most common form or pattern of the life-giving force on its journey. There would be every advantage in studying the individual forces in the equation of positive and negative, before categorically claiming the properties of one, while ignoring those of other energies at play, but they can be as subtle as they can be earth-shattering, so it is no easy task to establish quite what is at play.

It is not as hard and fast as science would have us believe and definitely not as simplistic. We can impose our theories to our heart's content, but Nature continues her simple way. It is perhaps possible for us to learn, but I suspect only by assuming our natural birthright: the role of observing.

PARA- AND DIAMAGNETISM

For scientific purposes, magnetism as it is manifest in our environment is considered for its properties as we saw above. However, the quality of force involved can be categorized into two useful and practical types of magnetism—namely, paramagnetism and diamagnetism—but the categories are not clear cut, as once more, there is apparently no hard line drawn in the sand.

Philip Callahan has perhaps been the most instructive voice in regard to this theory in recent years, drawing parallels in the natural world of insects and stone and connecting dots that have not been correlated before. In his useful volume *Paramagnetism*, he fundamentally explains that a substance (rock, soil, wood, whatever) is considered paramagnetic if it moves *toward* a magnet—yes, even wood has a mild magnetic capacity—and diamagnetism is simply a weaker form of magnetism. Water and most organic substances are considered diamagnetic.

Callahan is the only recorded modern scientist to have spent time investigating a round tower—specifically, the Devenish tower, near where he was based during the second world war. He even developed a very practical instrument—the Photonic Ionic Cloth Radio Amplifier—capable of measuring the energetic force in terms of the ELF and VLF frequencies emanating from a tower. Although he patented this device in 1993, it is no longer available as originally conceived, except in the extremely limited form of an earth-sampling device. His work was ground-breaking and invaluable for the modern researcher. He did, however, miss a few fundamentally essential elements with regard to the towers, but more on that later.

Once we move on from the limited "one or the other" way of considering what is happening, and integrate Nature's multidimensional system (which we cannot avoid) with her inimitable fashion of blending the two extremes, no matter what we call them—yin/yang, male/female, north/south, negative/positive, life/death—we can approach harmony. We are able to emulate Nature when we work in her service, along the lines of her patterns, and that is what, I would maintain, the ancients were doing with their stones, the pyramids, the stone circles and alignments, and of course, the round towers in Ireland.

MAGNETISM AND WATER

The relationship between magnetism and water is quite obscure, if not nonexistent from the modern scientific point of view, but when considered individually the connection might appear more obvious, as indeed it is when considered holistically. Given that our planet is made up of about 71 percent water, like the human body (what synchronicity!), it would be logical to work on the hypothesis that water is a mediating factor. Especially when we know that water is claimed to be the dedicated medium for the "magnetic force" coming from the moon and the essential component in the alchemy of vital force from all sources of life as we know it. The movement of water in the atmosphere, on the surface, and below ground makes it the prime mover of all the elements on which we depend for life here on this planet.

It would require several volumes to do justice to water—its development and its capacities and functioning, as well as its mythological references. One thing is quite apparent, however, and that is the awareness of the ancients concerning the importance of water and its abilities, as we can surmise from the vast pantheon of gods and goddesses of water, not only in the skies and on the earth but also underground. There is also a plethora of mythical beings related to those aspects of the combined effect of water and its surroundings, dimensions that were as strange to the ancient observers as they are for us now. A few examples of those beings would include the wyvern, naga, dragon, mermaid, and so on.

If it is unknown today precisely—or even vaguely for that matter—how the human reacts to low-field magnetic forces, one can project what might have been in the minds of ancient humans and the imagery they devised to deal with the inexplicable, but oh-so-real, forces that have not changed in manifestation since time immemorial, even though they are somewhat weaker now due to our incessant tampering with Nature.

What is more, the pre-Christian traditions in Europe and animist cultures elsewhere in the world all give importance to certain zones where water rises to the surface, such as springs and wells. Frequently, these sites were the subject of considerable local and more-distant attention, with pilgrimages being made to the sites that often became shrines, widely respected and protected by their reputation alone. These springs were, and still frequently are, venerated for their special healing aspects, qualities, or inspirational properties.

Naturally, whenever a new faith arrives on the scene every effort is made to eradicate the previous beliefs, especially in the Western world. Early Christian practice was extremely militant in doing this to the preceding traditions, especially the pagan Druids, Irish, Welsh, and Celts. It was soon apparent, however, that destroying the sites would cause an adverse and hostile reaction, so what better way to deal with it than to incorporate the former beliefs into the new religion, keeping all parties content as time worked slowly but surely to install the new faith.

The problem of course is that the raison d'être of the site is slowly forgotten, as are the technical details of the whys and wherefores of the techniques employed to build them. What we need to keep in mind, however, is that while the original purpose may have been forgotten, it has seemingly made no difference to the ability of the water at these sites to heal the visitors; the miracles still occur.

Consequent to the removal of the natural and practical reasoning for the original setup, all explanations as to the importance of underground currents are eclipsed and soon replaced by some article of faith, which completely blurs the understanding of something so simple as a geological fact. This then becomes the source of the "magic," which will still change with all the factors that affect hygrometry as we know it now. And that is something we tend to disregard, despite it being quite easy to prove with a few simple experiments.

In a manner of speaking, water stores energy, a fact as recognized today by authorities like Gerald Pollack as it was in ancient times in one of the oldest Vedic texts, the *Chandogya Upanisad*, which said that water is prana (life force or vital energy). We had to wait for Viktor Schauberger, however, to inform us that this is probably very dependent on the quality (or maturity) of the water. What a long way we are from the modern, sterile concept of H_2O!

There is a very tight relationship between the forces on the surface of the earth and those within the earth. We have a very limited nomenclature or definitions for these energy forms, and Viktor Schauberger is surely the only person in recent times (he started writing about his observations of Nature in 1929 or so) who has not only drawn attention to the actual function of these natural forces but has even tried to explain and connect them.

A lot can be deduced from Schauberger's unconventional and unorthodox ideas on water and its management thanks to his in-depth understanding of its functions. His foremost discovery was the role of water's temperature in its health, and he stressed the importance of the 39°F transition temperature, the temperature when water reaches its densest state. (Should you wish to delve deeper into this idea, you can consult the work of Martin Chaplin, Callum Coats, Konstantin Korotkov, Mae-Wan Ho, Gerald Pollack, and Vladimir Voeikov.) Schauberger would seem to be one of the few men to have expressed his concern for the superficial and ignorant way we treat water, and I share his apprehension. In my professional capacity as a technical translator in France for some twenty years, I worked for several large water companies and even had the chance to visit some of their installations. Hands-on experience of the technology employed was hugely instructive, but sometimes equally horrifying.

In all probability, the public authorities have no other solution but to add chlorine and fluoride to the water supply, but not only is that extremely damaging to human health and the environment, it complicates the purification process of water when it is recycled.

For example: Water is recycled when collected from the waste disposal facilities that equip most large conglomerates. It is then processed, treated, and redistributed into the main water supply system. In other words, water is consumed, urinated, recycled, and consumed once more. The purification methods generally used eliminate many of the nastier bugs but do little or nothing to the traces of estrogens, recreational drugs, or antibiotics that remain in the water *after* the various processes.

Release of the "purified" water occurs on successful completion in the final tank. A fistful of very small crustaceans is then thrown into the water. If these creatures scamper across the surface of the water, out the water goes into renewed service; if they expire, which they will do rapidly if the required conditions are not met, the water goes back into the purification process. A rather primitive but totally prevalent and standard procedure throughout the world.

This makes it easier to understand the purely chemical approach to water, as opposed to the holistic, life-giving function. If we are consuming a potentially regular dose of chemicals, already assimilated by other humans and animals, it would also provide the basis for meaningful research into many of society's modern ills. But that research is for another day, as are the effects on humans, animals, and agriculture, but it could go a long way to explaining the increasingly common complaint of intestinal fungal issues—the gut bacteria are no longer subject to the conditions essential for their existence, and the period of adaptation could potentially be long, if it is even possible.

Water is multidimensional in its scope of action and influence, with an intelligence of its own and memory being a component of that intelligence. As mentioned above, Jacques Benveniste lost his credibility in 1988 for stating something similar, but it seems that we are at last opening our eyes to this reality, with mainstream science even accepting Gerald Pollack's fascinating work *The Fourth Phase of Water*.

However, let us return to diamagnetism and paramagnetism. From a broad, holistic approach, as opposed to the atomic perspective adopted by science, the subject is perhaps worth considering by looking at what happens to water in the open seas. It evaporates, we are told, but we're not given any of the exact details on how. I would venture that surface water is less diamagnetic than the water at a lower level, most probably due to the temperature difference—the seminal 39°F as demonstrated by Schauberger—and is able to levitate (despite gravity!), attracted by the more paramagnetic environment found in the atmosphere. This is yet another example of the very weak field of magnetism playing an essential role in governing life. Nature is the most amazing manager, so I can only hope that the effluents from the land, laden as they are with chemicals and irradiated substances, prove amenable to her capacities out in the open seas.

WATER AND STONE

There are certain people, incidents, and things in our lives that leave indelible influences. The aforementioned Viktor Schauberger was one of those for me. His simplicity, perspicacity, and candidness are shining examples of what we could well emulate to understand nature. What is more, he shared his knowledge and the fruit of his observations freely. One particular incident that he recounted from his forest warden days in upper Austria is especially important to this narrative.

In a region of the high mountains where water was scarce was a spring covered by a dilapidated stone dome that Schauberger thought was in danger of collapsing. His older hunting buddies warned him of the danger of the spring drying up if the stones were removed. Attentive but skeptical as always, he had his team number the stones as they dismantled the structure. Sure enough, the spring ran dry, so he rebuilt the primitive structure in the precise order of its disassembly, and three days later the water bubbled forth again.

While we may not be able to fully explain the whys and wherefores of what occurred, there is obviously a strong argument for the use and application of stone when there is a scarcity of water, or, as in the case above, when there is a need to encourage water to the surface for human settlement or agriculture or the maintenance of the harmony that water in the right quantity ensures for all of Nature.

I mention this simple incident because it illustrates a major unknown relevant to this book—namely, the unsuspected properties of stone and its interaction with water. Schauberger's form of observation was holistic, focusing on water because he knew how important that element is to life. So he played with it and discovered insights into the intelligence and capacities of water that no one else has recorded to date.

Given that a substantial area of the earth is made up of water, it is likely that the component anions and cations in saltwater have a direct effect, due to conductivity or straightforward sharing ability, on their neighbor, rock. When we have taken a closer look at a rock to establish whether it has what I refer to as magnetic polarity, it has invariably proven to be the case.

This is a fact that few people are aware of: a stone/rock/boulder/menhir/dolmen/tower possesses magnetic polarity; hence the notions of male and female or yin and yang associated with stone. A simple nomenclature for sure but of significant relevance, given that we are surrounded by them. In addition, it appears that certain types of stone achieve substantially different forms of energetic output. This is significant because it is key to understanding all the erected stonework considered in this study.

The ability to arrange the components of a stone into a magnetic structure that can be recognized by us as such is in all probability due to the influence of water, or fluidity, and its ionic exchange and varying temperature. The idea here is not to investigate the scientific cause and effect, even if we were able to deduce the how and the why, but rather to discover what use can be made of this understanding to enhance our lives and benefit the natural environment, as there is strong evidence to believe that this is what the ancients did. They used stone to adjust the movement of energy to benefit one and all.

MAGNETIC POLARITY IN STONE

Some fifteen years ago, I was completely unaware of this strange magnetic capacity of stone. I was teaching at Mahachulalongkornrajavidyalaya University in Chiang Mai, established in Wat Suan Dok, a school exclusively for Theravada Buddhist monk students of humanities from Asia, mainly India, Nepal, Bangladesh, Burma, Thailand, Laos, Cambodia, Vietnam, and Korea. One day, at the end of an academic writing class, which no one could give a damn about, I was explaining the notions of Shankara's kevala-advaita philosophy to a group of mainly Cambodian students in their fourth and final year, when a pale-faced monk by the name of Phra Chris (PC for convenience) walked in and asked to join the class, which he did.

At the end of the session, he asked if he could join us the following week as he had mistaken my class for the one that he wanted to attend but found the subject much to his taste. PC was a recently ordained Polish monk living in Wat Umong at the foot of Doi Suthep, Chiang Mai's sacred mountain. For many years, he had lived on the island of Skye where he had been befriended by a German radiesthesist who taught him all about radiesthesia. This new-found acquaintance from among my Buddhist monk students agreed on an exchange of services: my knowledge of and experience with the laying on of hands and fasting for his knowledge of radiesthesia. And so started an intense, three months of experimentation and research.

We both shared a keen interest in sacred and applied geometry and alternative medicine and techniques, especially the practical applications that such arts permitted. He was an accomplished dowser, and he taught me—a ready and natural student—what he knew. Our main focus was on the use of geometric patterns and forms in our large garden in Mae Ann, just outside of Mae Rim, in the Himalayan foothills.

One day PC mentioned in passing that there was a large pile of rounded river rocks in Wat Umong and that the abbot of the monastery had given him permission to use them. He was of the opinion that river rocks had some special property and was keen to experiment with them. We collected thirty or so large stones, small boulders weighing 15 pounds or so, and brought them back home in my pick-up truck.

Using a pendulum and a Bovis scale chart (more on Bovis shortly) ranging from 0 to 40,000, we started recording our findings. Initially, we wrote down the score of the individual rocks to be used. Holding the chart over the stone in question and then asking "What is the energetic reading?" proved to be the surest way as the pendulum swung towards the relevant score. The same question would be asked when querying a given area, while adding "for this specific area." Then we placed them in the desired zone and recorded the reading on the scale again. We carefully measured out distances with a measuring tape and then placed the rocks in various geometric forms (squares, circles, and rectangles), including layouts incorporating values such as pi (3.14159) and phi (1.618). We were intrigued by the results and substantial distances that the influences of these patterns carried. There was, however, a strange phenomenon that was causing some confusion in our minds.

Even though we were using exactly the same measured distance when laying out an identical form, a square for example, we were getting readings that differed by several hundred units. Was it the ground (the local earth has a heavy component of metal) or the presence of trees or humans . . .? It was a tough nut, and of course, it was more by accident than logic that I discovered one day that the same rock was giving a different reading depending on which side was being measured. By simply turning the stone over, the reading was found to be substantially different, by a matter of a thousand points or so. We did not have any magnetometers, and even a meter that I later acquired did not give any indication of what we now knew was a subtle, occult property of a stone, so a pendulum and a chart it was.

Because a pendulum can only provide a binary answer—yes or no, positive or negative—it occurred to me to ask, "Is this positive?", not knowing quite what the connotations were to such a question. The side of the rock over which I held the pendulum produced an affirmative reply. I then turned the rock over and asked, "Is this negative?" I again received an affirmative answer. When I asked what the score was for each side, I received 15,500 for the negative side and 13,500 for the positive side.

Applying an apparently logical approach, we deduced that if the same rock gives two readings, one side higher than the other, the effects of the higher (negative) side could potentially produce higher aggregate scores from the whole layout. Would that provide empirical evidence, at least, that we were on the right track?

When we repeated the same patterns, but with all the rocks turned negative side up, the readings were off the Bovis chart. For those of you interested, you will find some recordings of our experiments in the appendix on Chiang Mai Stone Readings.

A word here about the Bovis chart might be in order. In 1930, André Bovis from Nice in France, a boilermaker by trade, devised a system to measure the vital energy of food, specifically the fruits and vegetables from his garden. He observed that their nutritional energetic value diminished over time from the moment they were picked or cut, especially the leafy vegetables. He created a scale, which he named the biometer, with a measurement spectrum ranging from 0 to 10,000, calibrated in units that he believed corresponded to angstroms, and then recorded his findings using the score indicated on the chart by a pendulum held in his hand. This scoring unit later became referred to as Bovis units in circles interested in physical radiesthesia.

Consequently, PC and I designed a new chart, this time from 0–100,000. We took new readings and found some very elevated scores, 45,000 to 60,000. And it turned out that spending time in such "high energy" environments clearly had an exhilarating effect on us, but it was only when my wife remarked that my hair was standing on end (PC, as a monk, had a shaven head) and both of us were especially bright-eyed, that we became aware of the altered state we were both in from working in the neighborhood of these rock formations we were creating in the garden.

In reviewing our findings, we came up with a number of propositions. We agreed that one of the special characteristics of a river rock is its magnetism. Perhaps it had assumed the round form as a result of being rolled in a river or of the water flowing over it for many years—perhaps even hundreds of thousands—which probably resulted in them picking up a charge, for the readings taken with the pendulum and Bovis chart differed from one rock to another, and rightly or wrongly, we assumed and called this the "magnetic charge."

There were a number of intriguing phenomena discovered as a result of setting up these "power patterns." The first was the shifting of a point where lightning was in the habit of striking. This spot beside our guest house was clearly defined by the lack of vegetation because of the regular scorching that the area of 13 square feet or so was subject to. Northern Thailand is not especially bothered by lightning although there are numerous storms in a year. However, some months after placing the stone circle, the strike zone had disappeared as the area grew grass anew, as it still does today.

The second event was the appearance of what I can only describe as arranged energy points, valid so long as some of the more powerful patterns were in position. In the process of checking readings at a distance from the center point, we measured distances in paces (feet) and at given angles. While doing that, it appeared that there were areas with higher readings regularly dispersed every 50 feet, seemingly in a well-defined pattern, running east to west. The individual point was showing a consistent reading of 20,000 over an extended area of 4.9 feet. That extended zone had its limits when a new telluric force (e.g., underground streams, caverns, or geological faults) manifested on the surface, or topographic incidents (e.g., lakes, slopes, and so on) occurred. Apparently, when one modifies the magnetic influence on the surface, by laying out strongly magnetic round river rocks for example, the neighboring unfolding materialization of magnetic fields is altered.

It might be timely here to mention an anecdote about using the magnetic properties in stone as a therapeutic remedy. Shortly after our discovery of the magnetic polarity of stone, I was called in to help a bedridden patient in very much a last resort type of thing in a desperate appeal for some small relief. Not knowing anything about the person or their condition but much concerned when confronted with such total weakness, I felt the situation required a careful application of a minimal form of energetic force.

I asked with the pendulum if there was any benefit to be gained from that quarter. The answer was affirmative, and two round river rocks later, one on either side of the bed—negative magnetic polarity facing up at the right for the positive charged hand, and positive magnetic polarity at the left for the negative charged hand—a gentle source of energy was introduced. In all probability, if a laying on of hands session had been performed instead, the patient would have passed, as they would have received far too much energy for their delicate condition. My hands, like anyone who practices these forms of therapy, come in with a score of 70 (for the left) and 65 (for the right). The rocks were at 35 and 30. The measurements were taken from a chart of my own design—the same as the one devised for the experiments with the rocks, since there is no need for a measuring unit, unless one is wedded to quantitative science. Comparison is a better standard, as it tells you immediately if there is alteration or not. Incidentally, the patient recovered and went his happy way.

What happened next for me was a period of intense study into all subjects related to magnetism, physics, radiesthesia, Egyptology, and what is rudely termed "alternative medicine" in a concerted effort to discover if there was further information to be found regarding this natural phenomenon and its possible use in the past. Over the next few years I read a great many books on these subjects, from the classics in the genre to the mavericks, and nowhere could I find anyone who shared this appreciation of stones having a magnetic charge or what could possibly be done with them, despite the ill-defined allusions to subtle energies.

However, my interest was seriously stimulated, for it is not only rocks that share this magnetic charge, but people, too, as I was about to discover from the works of Leon Eeman, Franz Anton Mesmer, and Karl von Reichenbach. It was not as straightforward as might appear because not only do the forces in our environment change, but they also have an intimate and immediate effect on our metabolism. We neither recognize that, nor actually see it happen, although we may suffer physically without making the correlation.

We will discuss the magnetism in stone in greater detail in chapter 6, but it might be helpful at this point to understand more about this connection between humans and their environment.

PERCEPTION OF THE HUMAN-ENVIRONMENT ENERGY EXCHANGE

We all—humans, animals, and plants—perceive things. Perhaps even inanimate objects do too; I wouldn't discount it. The majority of people are also aware of their surroundings. They perceive things with all 365 or so senses, not just with the five Aristotelian senses. Now while they may not be able to say exactly what it is they are feeling, they know if it is comfortable or not. If it is an influence that causes discomfort, they are sensible enough to remove themselves, when possible, from that troublesome source. What is more, sometimes they might even be able to determine the cause of the unease and change certain factors so that the problem is resolved, and harmony is restored.

Well-being, harmony, and comfort are expressions that convey what all beings on this earth are after. That much seems clear, and it would be reasonable to assume that has always been so in the known history of human existence.

Rather than engage in metaphysical speculation as to what this balance is and where it lies, can we simply assume, along with the Chinese, that when heaven and earth are in harmony, there is the environment suitable for human existence? A basic enough place to start our investigation that fits the bill to perfection.

The importance here lies in the fact that this sensation of health or well-being is facilitated, if not determined, by the synergy of what is happening in our environment, which of course, can only be brought on to the radar screen of our awareness thanks to our senses. A highly complex process no doubt, and one which depends entirely on the right balance and/or maintenance of temperature, air, water, light and movement, and such like. A very delicate act that can come to a grinding halt if any of the components are modified to excess or depletion. This has apparently always been the case, and even though there have been a few wipeouts in the past, we muddle along in those narrow parameters that separate life from extinction, oblivious of what is actually happening, let alone how it actually functions.

During those periods of muddling along, we humans devised and created some remarkable ways to improve our lot. The common denominator in the methods seems to be the interface and energy exchange between the individual body and the external world, with the emphasis generally placed on improving life for the individual by using elements from the external environment rather than the other way around. It is rare for humans to work for the benefit of the environment, but this is an idea that will be explored here because the unsung hero workers of stone seemed to be doing just that.

But what exactly do we know about how energy exchange really works? Rather than start a polemic that is unlikely to be resolved, can we assume a holistic approach, an overall, nondifferentiated, interconnected, and communicating view, as much as that is feasible for a simple human to achieve?

There is a dearth of information concerning this aspect of energy in our Western cultures, but that is not the case for other current and older cultures. If energy, as expressed in the living organism or the body, is understood as the vital force organizing the flow of blood and fluids, breathing, the nervous system, and the vital organs, all operating thanks to a constant flow, we not only find affinity with the theory of the meridians in Traditional Chinese Medicine, or the three *doshas* (*vata*, *pitta*, and *kapha*) in Ayurvedic medicine, but there is strong evidence that the ancient stonemasons had a firm understanding of the flow of energy in the earth. The common denominator is movement, no matter the speed. Energy can only act and interact when movement occurs. Death is the absence of movement.

Modern science has done a grand job of differentiation, determining the potential energy from the kinetic and introducing the six basic families of energy: thermal, chemical, electrical, mechanical, nuclear, and radiant. What is very encouraging is that there are even some schools of modern physics now coming round to the idea of an inter-connectedness of these different forms, either as a coherent totality or from the position of individual awareness. This is wonderful but some-what like reinventing the wheel for those familiar with thought systems from older civilizations; however, better late than never.

We moderns have still not yet grasped how energy "operates;" I would venture for the simple reason that the emphasis is constantly on the physically apparent manifestation. Little or no attention is given to the subtle processes that go on invisibly in those parts where we do not currently look. For example, in stone.

Energy, whether kinetic or potential, is organic; it is living, whether that is apparent to the senses or not. Everyone seems to agree with that, and you might add, "And so what?" Everything is formed of energy and is therefore alive. There can be no exceptions to that law. We just cannot understand with our reasoning capability what energy or life is. But what could be more natural? We are a part of that life. Our physical bodies are formed of materialized energy so to speak, and our spirit or vital force is energy personified, the aggregate of thought necessitating a physical entity to accomplish its desires. It would be inconceivable for the energetic entity of the human to some-how position the sense organs into observing and analyzing their own source by means of their own manifestations. But to have an intu-ition as to what is going on is another matter because we are no longer restrained by the trammels of the static cerebral intelligence that tend to be rather unmoving.

If we can put aside all considerations as to quite what we are as indi-vidual human beings and consider a broader picture where consciousness apparently animates a complex energetic mass that we neither control nor comprehend, then we stand a chance of catching a glimpse of real-ity. This compound of consciousness, intent, body, will, and especially thought seemingly differentiates us from the rest of "animal" existence, and that difference is probably what makes us think we can understand it, simply because we would like to.

We share the same energetic properties as plants and animals, but is it not this force that enables existence? Is that not reality in its most pristine form? Never to be perceived but always present?

Existence shares the characteristics of consciousness and well-being in the Upanishadic teachings. It would seem apparent, however, that this sense of well-being can only come from serenity, and as Lao-tzu stated back in the sixth century BCE, this is a state of mind available to one and all. Obviously only a few people practice that way of being, and as a result, it appears exceptional. We in the Western world have learned to differentiate psychology and medicine, whereas these two were the same thing for the ancient Chinese, for whom serenity was succinctly summarized as the control of emotions, or rather, not being disturbed by them. This is achieved by being constant, as it is for the Vedanta of the Hindus, and its accomplishment is the illumination or realization of what's going on.

From a modern materialistic viewpoint, constancy is considered something boring and dull, even to be avoided. That wasn't always the case. In the environment we know here on earth, there is nothing quite like stone to represent constancy, and it would appear that our forebears appreciated that the equilibrium needed to be in tune with our destiny was embodied in stone, and what is more, that equilibrium could be enhanced by the use of the inherent magnetism of stone.

The use of stone for human habitation is a relatively modern phenomenon, evolving at a rapid pace in the last few centuries from wood and earth to the predominant medium of concrete. The monumental structures found throughout the world no doubt served a variety of purposes, which are beyond the scope of this research, but generally speaking, were for the benefit of one and all in the immediate—and not so immediate—community. These structures were not used for individual accommodation.

That elusive balance, sought after by some, discarded by others, has in all probability not evolved over the ages, whereas it is more the lifestyle that decides the method and the means. That is what makes it difficult for someone of another time and space frame to comprehend the fundamental motivations of those from an earlier epoch. For all intents and purposes, modern life (especially city dwelling) employs an utterly different mindset from the one used by people required to live according to the dictates of Nature, not only with regard to lifestyle but to priorities as well. In the distant past the emphasis was perhaps more oriented toward unity rather than self-fulfillment, an association with Nature rather than random alienation and vicarious satisfaction.

Over the course of time, Nature has allowed/inspired humans to devise and create some remarkable ways to improve our lot. The problem for us today is to imagine what the purpose might have been in earlier times—presumably our ancestors were not just taken with a sudden urge to build stuff. This is the idea that will be explored here because the unsung hero workers of stone seemed to be doing something with a specific idea in their heads when they deployed them for the benefit of society.

As is often the case in human tradition, we either ignore or forget the original purpose of objects used in our daily lives. This seems to be the case with standing stones, whether menhirs from megalithic cultures in Europe and America, the lingam in presumed phallus worship of Hindu origin, the pyramids of Egypt; the round towers of Ireland and Britain; the *moai* of Easter Island; Hindu, Buddhist, or Khmer archeological vestiges; and the supposed territorial marking stones used in Southeast Asian architectural sites (called *singa* or *sima* in Pali, *sema* in Thai, *srnga* in Sanskrit). It is of etymological and intuitional interest to note that the Sanskrit word *srnga* means "horned" as in the horns of a cow, which act as antennae. This will be of interest in later discussions.

In our modern age, it is the results-based criteria that carry the day, so applying that same model to the subject here, I propose a closer look at the physical structures in question using a different means of analysis and understanding, which despite the current context of academic secrecy, logic would indicate that there must have been a clear purpose at the time of their construction.

Before we get there, it would be in the natural sequence of things to look at magnetism in the body first.

5

Magnetism in the Body

Water in the form of sap flows in a tree up during the daylight hours and back down during nighttime. Water, in the form of blood, flows through the veins and arteries of mammals, where it also follows diurnal and nocturnal patterns. Other vessels carry nerve impulses, and the meridians convey their energetic flows. These amazing processes occur for the duration of physical life, requiring no effort, nor cognizant control.

What is remarkable is that we pay no attention to the magnetic charge these flows generate because there can be no doubt that they do just that. The resultant magnetic aggregate is surely a major constituent of life and merits the epithet of biomagnetism. It is a significant health factor in frequency medicine and of special importance nowadays when our biotope is totally polluted by electromagnetic frequencies of the most unnatural sort. I would even venture, on the strength of my work experience, that maintaining a strong magnetic field is key to good health.

It is intriguing to find that this quality of magnetism is present even after the fall of the physical envelope, inasmuch as the magnetic charge or field of force continues its presence at a constant, albeit significantly lower rate than when living—namely, at a score of 15. This could perhaps indicate some form of remanence relating to existence, never lost but simply transformed. Whatever is going on is politely ignored by medical science—we measure the electric current of a patient to reveal reputedly important factors like heart function, but elect to sideline the associated information and remain in complete unawareness as to this aspect of life and its components. This idea is part and parcel of the theory being developed here, so please bear with me.

The magnetic, or biomagnetic, field of an individual can be an immensely useful indicator of human health and ranges in measurement from 15 to 75 on the chart of my fabrication. There is no unit of measurement per se and no registering of the electron spin or other movement.

The unit of measurement used to record the magnetic field of the earth is a microtesla, and the earth's charge ranges between 25 and 65 microteslas. I suspect that humans have a similar field with a comparable scale of measurement and is what I am picking up with a pendulum on my chart. Quite what this expression of movement is can be explained in a variety of ways, but I don't think that really helps our understanding. It is probably a function of the life energy, the Chinese qi or the Vedic prana. The strength of such an element in a living human varies according to many factors: exposure to electromagnetic radiation, the strength or otherwise of the immune system, the wearing of rubber-soled shoes, environmental influences, physical blockages, emotional well-being, and so on. In any case, it can be strengthened, and that is very good news because our entire environment is gradually being weakened magnetically.

It is useful to know the magnetic polarity of a person's hands if you are working on them in person, for it seems best to work on the positively charged hand. Generally speaking, men have a positive charge emanating from the palm of their right hand and a negative one from the left. Men who were born prematurely or who are gay often have an inverse polarity, making the left palm the positive and the right negative.

Women, in contrast, have a positive charge emanating from the palm of their left hand, although women born prematurely or who are lesbian may have inverse polarity. And almost invariably at menopause women's polarity switches, the right hand becoming positive and the left negative. This last observation might explain the severe changes to women's hormonal system, as experienced with hot flashes and other uncomfortable effects. It is not an instantaneous reversal but a gradual process, often taking several months, even years; hence the inconvenience and discomfort of menopause lasting in some cases for several years. The condition of being right- or left-handed, or even ambidextrous, seems to make little or no difference to the biomagnetic polarity of the palm of the hand.

The images below show the locations of general biomagnetic points in the human body. As you can see, there are a number of constant points shared by both sexes—the forehead is positive, the back of the head, negative; the top of the spine, positive, the base, negative. There is one neutral point—the navel. That is perhaps no surprise if you consider that it is the source of life from our mother. The birth of a child is a stage in the cycle of life, placed between the disintegration stage, death, and a new life; the arrival point, the navel or nabhi chakra (the only chakra mentioned in the Charaka Samhita, the ayurvedic classic), retains the zero-point polarity, specific to creation.

With the same constancy, the back of the hand will have the opposite polarity to the palm, irrespective of the polarity of the palm. If the left hand is positive, the right foot will be positive too and vice versa for the right hand and left foot.

The naturally harmonious, positive and negative arrangement of biomagnetic polarity in the human body is generally constant for the duration of a lifetime. However, there are circumstances in which certain aspects of the arrangement can change, most notably for women at menopause, as mentioned above. Other instances of polarity change may be a recent phenomenon caused by the surge in the past one hundred or so years of electricity, mobile telephony, wireless devices, personal

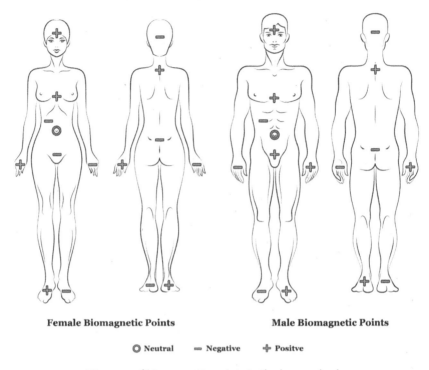

Female Biomagnetic Points   **Male Biomagnetic Points**

O Neutral   — Negative   ✛ Positve

Diagram of biomagnetic points in the human body.

computers and other electronic gear, air travel, the earth's weakening magnetic field (due to the extraction of oil and minerals), and so forth, likely resulting in the perturbed magnetic patterns and a plethora of modern diseases commonly encountered nowadays.

This fact is patently ignored by modern medicine and perhaps most other systems too; however, there is some relief to be found through magnetic recharging. This involves applying Leon Eeman's simple exercise that allows you to recharge your biomagnetic field in just a few minutes. Here's how: regardless of the polarity in your hands, interlace your fingers and let your hands rest on your stomach or chest. The polarities will be joined, negative and positive. Now, your feet: if you have right-handed positive polarity, put your left foot over your right ankle. Conversely, if you have left-handed positive polarity, put your right foot over your left ankle. You will find out after five minutes if you made the wrong choice and you will feel enervated, so change the position of your feet. It is a very subtle sensation, and quite gentle.

With hands and feet joined while you are sitting at your desk, on the bus, lying down, wherever, you have closed the biomagnetic loop of the body's field and are recharging. What does that do? As explained above, the current of the blood's movement through the veins and the energy through the meridians result in an electric charge, which can be measured in millivolts. Ten millivolts, and you are at death's door; 70 millivolts—and you will not get higher—you are fighting fit. Most people come in at 25 millivolts, which is weak, further indicating that the immune system is probably struggling.

No one, to my knowledge, has ever pointed out that this is just what you are doing when you are sitting cross-legged in meditation, one hand in the other, but no matter. One often sees old people in this position; maybe their bodies are reacting instinctively.

The immensely useful therapy of magnets, or their application, has been generally neglected, and as such there is no branch of medicine that is concerned with magnetism, so it is perhaps quite deliberately ignored. That is reason for celebration because it would most probably have been monetized and perverted. So once more, this technique offers a very useful way to take care of ourselves and consequently ensure a healthy level of biomagnetism.

This digression into human magnetism is done to introduce the idea of our interconnectedness with our environment and, more specifically, stone, which we might suspect, but do not act upon. I would postulate that one of the ancient purposes of stone and its placement, especially the more powerful standing stones, was to better human health, for our forebears discovered that there was a connection to and, even better, a readily available application for stone—namely, recharging magnetically. There are a variety of ways to accomplish this; for instance, by lying on a flat stone, standing against or between standing stones, and so on.

It is impossible to determine all the effects at work in our environment let alone which ones are impacting us because not only are our metabolisms different, but our magnetic fields are too. There are so many potential factors at play that it is positively daunting trying to calculate what is going on.

Fortunately, however, some people do pick up on what is going on around them and underground beneath their feet in terms of environmental influences, but not many know of the potential issues caused by those influences, and even fewer know that they can actually do something about them once they have discovered what they are. It is a matter of sensing; however, unless you are familiar with Nature's ways, there is very little chance that you will associate what you are feeling with what is going on below ground or the manner in which that energy form moves.

In the same way that heat and hydrogen rise, those energies ascend from below the ground and head upward to the sky, no doubt in the permanent process of what we know as life—the recycling/regenerating effort that is in constant movement all around and within us. To gain a better understanding of these forces, we would do well to pay closer attention to the archaeological vestiges left behind by our ancestors and take note of the observations of past researchers from the Chinese and Indian cultures and traditions along with numerous others. There are in all probability messages left to us by our ancestors that have been deliberately occulted in man's never-ending effort to impose control over his fellow man.

Over the past couple of hundred years, several very smart people have expressed some remarkable ideas on the subject of magnetism, as is only to be expected from normally curious individuals attempting to perceive the workings of the subtle life force. These serious experimenters ranging from the eighteenth to the twentieth century had the following different and somewhat inconsistent things to say regarding human magnetic polarity:

- Franz Anton Mesmer (1734–1815) observed that all points on the left side of the body have the opposite polarity to those same points on the right side. He believed that the poles can be changed, communicated, destroyed, and intensified.

- Baron Karl von Reichenbach (1788–1869) found that the left hand has negative polarity, as does the od (vital force) in that hand, in both men and women.
- Hector Durville (1849–1923) maintained that negative polarity is to be found on the left side of the body for both men and women.
- Leon E. Eeman (1889–1958) found that both men and women had negative polarity on the left side unless they were left-handed, in which case the polarity was reversed.

From this evidence, something bigger is clearly taking place, but no one in the scientific community—who might have the wherewithal to conduct thorough research—seems overly keen to investigate further. It is a measurable fact that the earth's magnetic field is weakening, and given the magnetic polarity of humans, it would be logical to presume that this is having a subsequent impact them. Would this not make human magnetic polarity an area worthy of further study?

If we are to find answers to these life-changing issues, it is essential to structure our research because it is apparent that the different researchers above did not ask the same questions during their respective experiments. In all probability, no questions regarding polarity were asked, and it is only too natural to take an accepted fact for granted, even if that fact is incomplete.

In my practice working with humans, I systematically question the following:

- What is the magnetic polarity of each hand?
- What is the individual's magnetic charge (on my scale)?
- What is the individual's electric charge (in millivolts)?
- What is the individual's degree of vitality and general health (a suggested scale of 0 to 10)?
- What is the individual's emotional/mental state?

Obviously, this is not an exhaustive list, and is very specific to my knowledge and understanding, but you get the idea of what could help ensure more coherent and consistent data. In all probability, the early researchers didn't ask their subjects any questions at all because they were simply not familiar with the possibility that these factors can vary, and they were perhaps even less familiar with modern scientific methodology, even if they believed there might have been some merit in interviewing their subjects to gather as much data as possible.

Confusion over the centuries concerning magnetic polarity in the individual's hand could have been avoided if each person had been checked before, during, and after the experiment. Because our understanding is deficient in these matters, as is only natural if we do not recognize the importance of biomagnetism in the individual, it could be beneficial to revise our way of looking at things, which, as we know, develops, evolves, and sometimes regresses.

Clearly we can draw no firm conclusions from these past experiments because of the lack of recorded parameters, but in my experience it is definitely worthwhile asking on each occasion, with a pendulum or some type of muscle testing method, the questions above. The causes, as always, are beyond our ability to explain with any confidence, but one thing is for sure: if you don't ask, you are unlikely to realize that you don't know.

It is important to be careful not to fall into the habit of dogmatically maintaining that a given problem can be worked out in precisely a certain way, as is so often the case today, with endless clichés about one practice being the "best" and another the "worst," don't eat this, eat that; avoid this, embrace that. It would be naïve and even impossible, even for an expert in the relevant field, to hold that a given practice is unequivocally better than another. Each individual will respond differently to the same set of conditions. If we all had the same metabolism, allopathic medicine would work across the board, and good health would be guaranteed for all, but unfortunately that's not the case.

Yet for some reason we tend to accept solutions without question if we think we're getting a fair deal—we convince ourselves that we can benefit, if only we accept certain conditions. But it is those conditions that are debatable that might not hold up under rigorous intellectual reasoning. In our haste to find a solution, we've stripped intuition out of the equation, and it's only in our heart's intellect that we can hope to find a true solution. It is our heart that has a connection with what is, while our head merely dictates. We know where pure rationality takes us and how easily it can be manipulated to the advantage of a few.

SPIRALING ENERGY IN HUMANS

Some fascinating research was conducted in France during the 1930s by Jacqueline Chantereine and Dr. Camille Savoire. They suggested that humans receive telluric and cosmic energy into the body along well-established pathways. Their conclusions led me to experiment further along the lines of biomagnetism.

Chantereine found that the telluric yin (negative) energy enters the body through the left foot. She found that the energy rises in a spiral through the body and exits via the hypophysis, or pineal gland, radiating through the temples and the eyes. As mentioned above, this constant in human biomagnetism resides in the spine and head and applies to men and women alike. Systematically, the front of the head (forehead) is positive, which is probably due to the energetic flow of the telluric force streaming outward at the level of the eyes via the pineal gland and the cosmic frequencies coming in through the top of the head to access the pituitary gland before spreading down through the body. The back of the head is negative.

The cosmic *yang* (positive) energy enters through the top of the skull, I believe via the pituitary gland, and spirals down through the body exiting through the right foot.

The figure on p. 86 demonstrates this trace of the spirals according to their research. I question the systematic configuration for both male

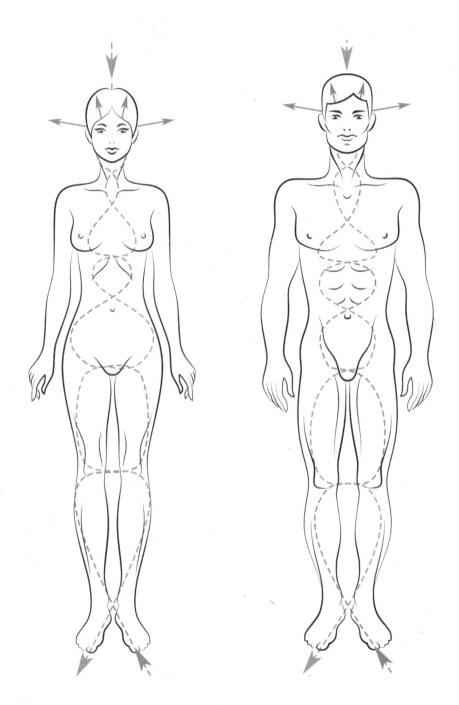

Cosmic and telluric force entry and exit points.

and female and tend to believe that it is an individual affair. The following is an excerpt from chapter 6, "Man's Electro-Magnetic Dispersion Field" of their book, *Ondes et radiations humaines*, written in 1932 and translated by me from the French:

> The dispersion field or overall vibratory system of any human, considered in its totality, includes the following manifestations:
>
> 1. A vortex departing the heart, rising in a spiral to the right, penetrating the frontal parietal suture, describing a descending circulation in a closed trajectory (in morphology it is known that descending circulation is the characteristic of vortices), moving to the left of the body's central line to return to the heart. We have given this vortex the name "individual vortex," because it vibrates in a given color range specific to each person; it is the essence of each individual, so to speak. When a person dies, this individual vortex persists whereas all other manifestations expire; we leave this scientific observation to the reader's philosophical contemplation.
>
> 2. Two vortices imprison the body; one rises from the feet to the torso, presenting the characteristics of the telluric force with ascending revolving cyclones. The other descends from the head to the lower body, presenting the characteristics of the cosmic force. These two vortices are not always in perfect harmony. When the telluric force predominates, passion is to the fore and physiological life more intense. When the cosmic force is dominant, the temperament is more cerebral, tending toward the spiritual.
>
> 3. An egg-shape form envelops the body from the top of the head to the toes, encircling the shoulders; it has a positive electrical charge in the right upper and left lower sections of the body, whereas there is a negative electrical charge in the lower right and upper left parts of the body.

4. There are two cylinder-shape force fields one within the other, with opposing polarities, the external one being positive for men and negative for women and the inner cylinder, negative for men and positive for women. In some individuals both cylinders are of negative polarity, as is their temperament. These fields are especially strong in newborns.

5. Spectral bands and rays, detected with our colored syntonisers, form what we call the "human spectrum." Figuring out these bands was our first task in the study with dark rays indicating damaged organs. These are the same phenomena that doctor [Walter J.] Kilner perceived using dicyanin screens which he then named "The Human Aura."

If one considers the idea of a vortex being the source pattern of flow in the physical world as we know it, then the "individual vortex" mentioned in the first paragraph could correspond to the notion of a personal pattern, somewhat like the astrological notion of the conjunction of stars and planets at birth or conception, affecting us throughout our terrestrial lives. If this pattern has a specific frequency, which is a reasonable enough assumption, one can further infer that such frequencies have ranges, which can be defined as colors, especially when referring to human health analysis with the idea of determining an individual's makeup.

This notion of cosmo-telluric energies flowing into the human body is also common in the ancient Chinese philosophy of Taoism, the *yin* energy coming from the earth, and the *yang* from the sky. As is often the case especially in ancient traditions, careful attention must be paid to the context of the terminology as the same word might be used in different contexts with which we are not necessarily culturally familiar. This energy consists of a range of what we choose to categorize but which probably cannot be separated in nature. For example, conduction, electricity, gravity, heat, induction, levity, light, magnetism, and sound are scientific concepts that we also do not fully understand but that sound quite familiar.

There does not seem to be any basis for reading more than a systematic pattern at work in the flow of energy here, and there seems no doubt that all physical forms, whether animal (including human), mineral, or vegetable, have what we can call a "biomagnetic" field.

Modern environmental conditions have changed so dramatically in the past few decades, with widespread electromagnetic and nuclear radiation prevalent in almost all countries of the world, that the earth's already weakened magnetic field has become even weaker. The ceaseless extraction of oil and other minerals and a weakening resistance in the ionosphere to incoming cosmic energy has inevitably had an impact on our biomagnetic potential. Ignoring these conditions can have unfortunate consequences, such as rapid depletion of energy, slow recovery from illness, fatigue, and even sickness, so it's a good idea to take the necessary precautions if you want to avoid exhaustion, especially if you practice energy work.

ELECTROMAGNETIC FIELDS

The scientific consensus would have it that our world is an affair of frequency, a region populated by infrared radiation and electromagnetic fields (EMFs), which have far-reaching biological, chemical, physical, and environmental effects. What's more, with regard to humans, EMFs are influential and sometimes determinative in coordinating and integrating our sensory, nervous, and endocrine systems. In simpler terms, we live in an invisible frequency soup and we're not quite sure what it does to what or to whom, nor when, where, or how it does it, but there is growing evidence that there is some form of relationship at work that affects our world, although not necessarily in a comparable way at the various structural levels. This soup might interest you because you are one of the ingredients.

One aspect of the electromagnetic field that cannot be explained by conventional theory is that it acts at a distance, a fact that is perfectly manifest in the numerous psi phenomena (thanks to Don Hotson for pointing out this scientific detail). Furthermore, EMFs operate instantaneously—just like thought.

Whenever something moves, it creates a flow of energy. The circulation of oxygen and blood that results when air is displaced while breathing, the solar wind blasting past our planet, the creation of water from oxygen and hydrogen under the surface of the earth are ready examples of this.

It is all very well to measure these causes, recording the fascinating details and feeling content with that knowledge, but what do we know of the process? Unfortunately, we are satisfied to accept the details once they are measured as somehow being the effect and leave it at that. While they may be true, they don't make up the *whole* truth, only a very small part of it.

We systematically ignore the forms and patterns assumed by energy in its movement. These flows, as we have just seen, are essential to life. Of course, this seems so obvious that it is hardly worth mentioning, but the fact remains that movement is the agent of change, and those forms are the patterns of life from which we can learn a great deal if we simply make the effort to observe them. We ignore them at our peril, for if we really want to understand how we can work harmoniously with Nature, we need to learn her ways rather than imposing our ways on her, which is ultimately impossible even if we vainly believe otherwise.

It matters little what we call whatever it is that flows in, on, or with these movements; we use them on a regular basis, and we can even observe this energetic flow at work in the environment, affected and causing effect when ambient conditions change. This is the origin of climate, which is the driving force behind all the products of Nature.

What's more, and not so generally recognized, is that these flows can act as a carrier for other frequencies, generating even further interaction and/or force fields. The actions of these flows may—or may not—depend on the strength of the original flow or on their interaction with other materials or flows. In other words, it's a total mess if you're trying to figure out how it works and attribute a specific value to a given factor.

But what do you observe? The evidence shows that our entire biosphere is alive, thanks to a remarkably well-organized system not only in which we live but that lives in us. That same system continues to maintain everything, even if we behave abominably toward it, showing no consideration or understanding. The arrangement either has immense forbearance, or it can handle whatever we throw at it. Hopefully, some of both.

Science teaches us that, historically, these flows in the troposphere (where we live) have been arranged and governed by Nature along electromagnetic frequency patterns established over several billion years. Science, however, is not so eloquent about the danger to the health of both humans and the environment that can potentially be partly or entirely attributed to a rupture in these flows; for example circadian rhythms, electric and magnetic fields, protein and water cycles, and other factors essential for the natural continuation of the world as we know it. So while there seems little doubt that these rhythms are being disturbed, no answers will be forthcoming unless we look more closely at the consequences of the enhanced levels of nuclear and EMF radiation that have recently been created. In all likelihood an answer can be found in Nature, but claiming that we need more time to research it is disingenuous.

A recent June 2014 abstract of a medical paper, "60 Hz Electric Field Changes the Membrane Potential during Burst Phase in Pancreatic β-Cells: In Silico Analysis," by G. F. Neves et al., states the following:

The production, distribution and use of electricity can generate low frequency electric and magnetic fields (50–60 Hz). Considering that some studies showed adverse effects on pancreatic β-cells exposed to these fields; the present study aimed to analyze the effects of 60 Hz electric fields on membrane potential during the silent and burst phases in pancreatic β-cells using a mathematical model. Sinusoidal

60 Hz electric fields with amplitude ranging from 0.5 to 4 millivolts were applied on pancreatic β-cells model. The sinusoidal electric field changed burst duration, inter-burst intervals (silent phase) and spike sizes. The parameters above presented dose-dependent response with the voltage amplitude applied. In conclusion, theoretical analyses showed that a 60 Hz electric field with low amplitudes changes the membrane potential in pancreatic β-cells.

The medical annals are replete with such papers on subjects related to the pathological impact of our modern conveniences, but contributions mentioning the potential dangers are few because there is no one to fund the research—surely the national utilities companies won't encourage such inquiry! It is so much easier to bury one's head in the sand and ignore the bad news.

Not that I am such a Luddite as to suggest abolishing electricity, mobile phones, microwave ovens, and other conveniences, but a little truth as to their effects on Nature or even just on ordinary men and women would do no harm. Or are we obliged to be complicit in the cover-up? My grain of salt comes in the form of the solutions that Nature has enabled me to discover. As Henry Hobart said, *"A non posse ad non esse sequitur argumentum necessario negative, licit non affirmative"* (If a thing be not possible, an argument in the negative may be deduced, namely that it has no existence; but an argument in the affirmative cannot be deduced, namely, that if a thing is possible it is in existence). Let's move on now from the more theoretical aspects to the more subtle and secret facet of stone.

6
Magnetism in Stone

Stones have apparently been around forever. They are several billion years old we are told. We use or abuse them, much as we do the earth's resources, without any compunction or second thought and with—at best—partial awareness of their properties. We know of their hardness, weight, density, appearance, strength, texture, porosity, resistance, and so on, and we apply those qualities carefully as a function of the intended use. However, few people—including stonemasons—are aware that stones have both a positive and negative side or end, as discussed in detail in chapter 4.

In the absence of another measuring instrument, take a pendulum in hand and suspend it a few inches over the rock. If you keep your mind free of thoughts and don't ask any questions, the pendulum will start gyrating, clockwise for the positive magnetic polarity, counterclockwise for the negative.

Perhaps the most convincing way of witnessing this is by engraving a stone. A number of years ago, I went to see a laser engraver to seek his help in engraving some geometric patterns of my design on wood. Upon entering his shop, I noticed that he was preoccupied and asked if there was a problem. "A big one," he told me and showed me two samples of engraving that had just been completed by his laser technician. They could not understand why, even with the same settings on the laser machine for depth of cutting, intensity, and time, the outcome was totally different between the two stones. One was very clearly defined, and the image of the 9 mm Eagle pistol leapt out of the rock. The image on the other stone, however, could barely be seen; there was no relief to the image, and it would not be acceptable to the gun club that placed the order for fifty identical pieces.

I asked if I could examine the two stones, using a pendulum to check for polarity, and he agreed. The clear image turned out to be on the negative side of the rock, and the hard-to-see picture was on the positive side. So I arranged all the rocks on the table negative side up, and they were then engraved. The client was delighted with the consistently clear-cut image on all the stones.

By way of demonstration, and for you to test your dowsing skills, here is a photo of a stone that you are invited to test to see whether the side facing upward is positive or negative.*

Round rock suitable for a polarity test.

---

*Answer: Negative magnetic polarity facing upward with a score of 70 on my chart.

You could also try this yourself with a round rock like the one pictured. Find a suitable rock and determine the polarity of each side. Then make a chart with a scale of −60 to +120 inscribed on it, and holding the chart in one hand over the positive side and the pendulum in the other hand, ask what the score is for the stone. Having noted the figure, turn the stone over and repeat the same process. The score will be higher for the negative face of the rock. In other words, the negative face of a stone has a higher magnetic field of force than the positive side. If one is dealing with paramagnetic stone, one could expect to find a score of 5 percent higher or more than the positive side. Every stone, even if it is diamagnetic, will still show this positive/negative polarity, albeit with a lower strength than a paramagnetic rock.

The scale of the chart could represent the centimeter-gram-second (CGS) system as proposed by Carl Gauss in 1832, later changed to a *gauss* as a measurement of magnetic induction. Basically, this is a measurement of how long one gram of substance, one centimeter from a magnet, takes in seconds to move to the magnet. In actual fact, the CGS scale is probably much lower in absolute values because magnetism is not an absolute; it is a function of the environment—a purely relative value yet a totally real and permanent influence in our troposphere.

The magnetic field on the surface of the earth ranges apparently from 0.25 to 0.65 gauss, as a function of the telluric *and* cosmic influences, not just the telluric ones as some would have us believe. There are most probably other influences also at work.

Having said that, please be aware that while this magnetic field is very weak compared to the field coming from an electromagnetic impulse (mobile telephony), electric power line, or even an electric cable in a house, it is powerful enough to influence the growth of plants and, of course, the state of health in humans. We are told that this effect is on a very subtle level; there are few studies to indicate, however, the range or degree, as a result we might do best to conduct our own research if we plan on protecting ourselves.

We can reasonably assume that this influence has been present over the ages, unlike the electronic devices of the past sixty or seventy years, although they are of great consequence due to the intensity of their power.

THE ELEMENTS

From what we can learn from ancient writings and are readily able to imagine, the elements were subjects of great concern for our ancestors. Weather is still the most common subject of conversation throughout the world, in all cultures without exception—an indication of its importance to our well-being. For instance, thunder was the subject of one of the earliest Indian poems in the Rig Veda; hymns 32 and 80 were made to the foremost deity Indra, who is associated with lightning, storms, and thunder.

Human concerns are very limited, and climate, weather, and other planetary influences are the cosmic forces about which we can do nothing, save perhaps apply geoengineering. We are subject to their unyielding power. However, if we are studious and observant enough, we can learn to work with them and figure out how to make our immediate environment more comfortable—whether it is to hide from the extremes of climate, find a suitable source of food, or simply to feel good. We can attempt to understand them, as we are doing here, so as to improve our lot, as the ancients were so clearly doing. The most complete legacy existing is to be found in Chinese texts, where we will now turn.

Chinese cosmology, as explained in their ancient texts, is full of symbolism and imagery and not always easy for a Western mind to grasp. But by using a little imagination and juxtaposing phenomenal events with experienced reality, it soon falls into place.

The philosophy is based on the five elements—earth, wood, fire, metal, and water—which are best considered as four domains around a center to which the energies come from heaven. These energies are then manifest on earth, and there is an "official" in charge of each aspect. The earth (harmony) is the center, where man resides, and the remaining four elements and their corresponding seasons—wood (spring), fire (summer), metal (autumn), water (winter)—assume their role therein. The names of these elements are taken from observed physical phenomena, and the characteristics attributed to each of them are, of course, based on the qualities of the specific phenomenon. But these five elements, so important to the Chinese way of being, are not substances as such, but forms of energy that inform every single substance and process of transformation. They are living, evolving entities. When expressed this way, with the aim of explaining a metaphysical (beyond the material) force with which we are all familiar, we can gradually perceive them as being the play of heaven and earth with humans caught in the fallout between them.

Of the two—heaven and earth—we are more concerned here with the earth, the telluric component, for the simple reason that is where we can modify certain aspects of these energies and influences.

TELLURIC FORCES

Telluric is the term used to define what comes from the earth—the yin energies. While these forces are inherent in the earth, they can sometimes be found to be more than uncomfortable to experience, especially when they derive from certain causes, such as geological formations, faults, underground streams, the crossing of waterways, underground caverns, mineral deposits, and the like. These will be discussed in more detail below.

What seems to be happening from the point of view of physics is that an individual geological formation can, and generally does, generate a movement that has its own very low-voltage electric current. That current has a companion magnetic field, and both the current and the field interact with the surrounding environment, knocking on up to the surface where they are released from the constraints surrounding them below ground. These forces apparently continue rising until such time as the incoming cosmic forces overcome and neutralize them, so to speak, which happens at high altitude. Fundamentally, the movement and the path they take will always remain a mystery. We can only guess at the pattern as a function of the effects, but it probably further adds to the complex of electric and magnetic fields of force.

The following are some of the most common telluric events:

- Underground water: When water flows through any underground passage, capillary, pipe, fracture, or fissure, the result is a magnetic field that can be surprisingly high in frequency and strong in effect. The field fluctuates depending on multiple factors, including what the water is conveying, the speed at which it flows, and/or whether it interacts with any other earth energy. The earth's natural energy field is particularly influenced by the crossing of two or more watercourses or by the crossing of a watercourse and another type of energy line. The crossing of two underground waterways, irrespective of their individual depths, is the most dangerous phenomenon for humans with regard to biological damage due to the intensity of the combined magnetic influences that occurs when they reach the surface.

- Geological formations: With their different mineral and magnetic components, geological formations are responsible for a lot of mischief. Their ability to allow or change the flow or directional thrust of the subterranean minerals or water is their main deleterious effect, but deposits of magnetically or gas-charged rock are also to be contended with.

- Geological faults: Caused by movement in the tectonic plates or the upheavals the earth has experienced, geological faults also affect the energy field at the surface. While they are less common, mainly because they do not often present zones suitable for construction, they can be very harmful.

- Underground caverns: Due to the resonance developed by the substances collecting or passing through the cavern zone, underground caverns cause problems for the inhabitants above. Invisible from the earth's surface, these holes are constantly creating energetic movement due to pressure, interacting with surrounding formations of a chemical and/or physical nature, and the natural passage of the effluents upward to the zone of least resistance— where we reside.

It is extremely hard if not impossible to determine the precise nature and origin of a combined field of force, let alone its evolution, as not only is there a lack of available instrumentation outside of the laboratory to measure and gauge the energetic strength of that force, but it seems to be further modified by additional and highly random factors, familiar to Traditional Chinese Medicine practitioners, such as time of day, atmospheric conditions of temperature and pressure, planetary influences, and subterranean determinants, which are not categorized. Even then, there is a singular lack of instrumentation sufficiently refined to measure the fine degree of fluctuation in movement or its impact. This is where the experienced radiesthesist—versed in the subject matter of frequencies and their possible interactions, equipped with the necessary charts and suitable tools, and sufficiently practiced to ask questions of the most diverse kind—comes into use.

While these frequencies may be tiny, representing a minute force, it does not mean they are innocuous and without effect. An effect on what, to what extent, and how are the questions, but they *are* going to have an effect. Hence the need for us to assume a much broader approach than the current cause and effect, force vs. power, materialistic view generally shared by humanity.

These geological events have their own footprint, so to speak, when the energy they produce reaches the surface of the earth and interacts with whatever is present there. As these events evolve, it naturally becomes quite difficult to appreciate and assess their effect because we do not recognize their impact—once again lacking the instruments to measure their effect on the human immune system or even the organs— nor do we record statistics in their regard. What is more, as these energy forms move they constantly interact with each other to varying degrees. As an example, the energy field developed using the system of wooden rods that I use to offset the telluric influences causing geopathic stress often travels extensive distances—four to five kilometers—until another stronger energy source is encountered and either weakens or strengthens the first. No rules here.

GEOPATHIC STRESS

These telluric forces are at work round the clock. When they adversely affect humans and animals, which they often do, they contribute to and form one aspect of what we commonly refer to as geopathic stress (GS). The magnetic component is especially relevant to us earth dwellers because those frequencies coming from the ground rise up to the surface where we hang out and can work serious harm. It is nigh on impossible to determine all the unitary effects at work in our environment let alone which ones are impacting us because not only are our metabolisms different, but our magnetic fields are too, changing with our state of health, diet, and subjectivity to planetary influence. There are so many potential factors at play that it is positively daunting to try to calculate what is going on.

However, when general health problems (physical, mental, emotional, or physiological) arise, however, we are keen to find causes, and if these energy forms are part of the equation, which indeed they are, it might be easier not only to explain the source but, far more importantly, find a solution. These low-field magnetic influences, so neglected by just about every branch of modern "science," have so much to account for. Going back in time to China and the earliest uses of feng shui, we find the recognition not only of the dangers of underground water because of what it does to people but also and far more pertinent, how those dangers affect the movement of energy inside the body as it flows (or stagnates) in the meridians, as recognized in Traditional Chinese Medicine.

For most forms of life and its normal development on earth, a balance in natural energetic vibration—or for the human and animal, resonance in the nerve and endocrine systems—is required. So for every living cell—whether human, animal, plant, or mineral—to grow, develop, mature, age, and finally die from old age and wear rather than from disease, it must, mandatorily, vibrate harmoniously. Therefore, GS can be seen as a component of disharmony because of the frequential disturbance it causes, possibly making things worse but definitely not reducing aggravation.

The influences of GS depend on its source type or origin, which are numerous and both natural and man-made and include geology, geomagnetism, electricity, electromagnetism, cosmic forces, shape and form, meteorology, and their combination and periodicity to name just a few. The effects on humans can be transferred and/or transported, be slow or rapid, and are generally cumulative over time. The World Health Organization estimates that 30 percent of buildings or structures are subject to geopathic stress and have what is known as sick building syndrome. I would put that closer to 60 percent from my experience over the past twenty years. These effects can also be further complicated (for humans, at least) by factors such as metals (beds and chairs), chemicals, poisons and pollutants, malnutrition/diet, water quality, recreational drugs, alcohol, psychosocial stresses, genetic weakness, electrosmog, chemtrails, and so on.

Basically, these energies create frequencies that interact with the environment and the people who are in the vicinity at their point of exit and represent a potential threat. Obviously, but not necessarily so, those people with a weakened immune system will be affected more than robust individuals, and these frequencies will further weaken the immune system as they are simply not conducive to a harmonious restoration at night when mitosis takes place.

Like humans, most large animals, such as cows, horses, and dogs, prefer to live in the absence of those telluric influences. Insects, ants, and other crawling bugs, on the other hand, are happy there—as are cats. Perhaps cats were sacred to the ancient Egyptians because they seem to have an unusual ability to handle geopathic stress quite well. Do they transform it to something beneficial? What is the story here?

Those people, animals, plants, and objects regularly spending time in zones subject to GS are likely to experience problems at some stage in their existence. GS does not cause illness per se, but it can definitely lower immunity, and it is constantly interfering with and altering the frequencies we depend upon for our material existence and can exacerbate health issues. Having said that, the strength of the individual's immune system is all-important, as those people who sleep over some telluric influence but have a healthy immune system experience fewer difficulties.

The health problems affected are varied, but they appear to cover a vast physical, emotional, and mental range, including addiction, ADHD, aggressiveness, allergies, anorexia, anxiety, arthritis, asthma, bed-wetting, bulimia, cancer, candidiasis, cell deregulation and growth issues, chronic fatigue, cot death (SIDS), dementia, depression, diabetes, eczema, enzyme production, exhaustion, fetal development, food intolerance, glandular fever, headache, heart disease, hyperactivity, infertility, insomnia, intestinal disorders, lethargy, lymph problems, memory loss, miscarriage, multiple sclerosis, myalgic encephalomyelitis, nervous disorders, obsession, pain, Parkinson's, premature birth, resistance to treatment, restless sleep, rheumatism, schizophrenia, sexual abuse, skin problems, stressed relationships in partnerships, stroke, tuberculosis, and on and on.

Initially for humans, GS often manifests as a sleep disorder, which is perhaps explained by the fact that mitosis, or cell replication, occurs during sleep. If a person sleeps in a GS-free zone, the cells develop in harmony with the original state of the cell. However, if the ambient surrounding frequency stops the cell from vibrating at the desired level, the cell becomes deformed in relation to the original. Imbalance and disharmony follow.

GS in humans, as well as telluric influences from the ground, can be easily detected with a pendulum and an adequate radiesthesist, although very often the sensitive human will feel a variety of sensations as a function of the nature of that force. The first thing to do on discovering the presence of GS is to remove it, for a person who exists in the influence of that condition cannot get better as long as they stay there. On the contrary, there is every chance their health will deteriorate.

Bed and work are where most people spend the majority of their time. While it is sometimes sufficient to avoid the danger by moving the sleeping area or desk a few feet, that may not always be possible. Placing a copper or aluminum sheet under the mattress will reflect the radiating waves from water, for example, when an underground stream passes below, but that will have a contrary effect if there is a geological fault, which could accentuate the harmful effects.

The best solution is invariably to use a harmonizing system as it can further usefully offset the effects of the other components that complicate our lives, such as radio masts, high voltage power lines, industrial processes, domestic electric appliances, and the numerous sources of daily exposure to magnetic and electromagnetic fields.

As explained further on, one of the probable purposes of the towers was environmental well-being. Presumably because there was either a need for that, or simply because it is possible, and as it is beneficial for one and all, let's do it! Nowadays, there are numerous devices on the market promising relief to the ills caused by modern gadgetry, and surely the easiest way to find out if one of these systems is the right solution for you would be to question with your pendulum. However, as traditional wisdom is my preferred go-to source, it seems more natural to use a system that is found—if you know where to look—in antiquity. One of the secrets revealed when researching the Chartres cathedral was the application of two measurements, specific to geographic location, subsequently used in the architectural design of the edifice and systematically employed by the Templar-architects in the cathedrals elsewhere in Europe. The system I use to enhance environmental frequency, and well-being for all within that space, is based on that know-how using eight wooden rods cut to very specific lengths.*

---

*This service is readily available for a modest fee on my website, radiesthesia.online.

The effects of GS do not only concern human health. Houses with GS running through them are consistently slower to sell than those without (it would appear that people sense something's wrong even if they are not consciously aware of it), lightning strikes affected areas on a regular basis, and GS is often a factor in struggling businesses, accident black spots, failing technology, corruption, financial decay, and bad luck in all its forms.

To my knowledge no scientific biomedical journals have published any articles in the English language concerning GS. Having said that, it would appear that since around 1900 doctors have recognized GS as a pertinent factor. A medical doctor in Germany, Gustav Freiherr von Pohl, found that a number of his cancer patients all came from the same village. When he asked a dowser to go and check the conditions in the village, the latter found that all the patients were subject to GS in the form of an underground flow of water over which all of their homes were located along one side of the street, and in addition they were all sleeping in the zone of influence.

Since that time, there have been a few doctors, mostly involved in the alternative fields, who have paid attention to this phenomenon, but far too few to make the necessary impact of awareness on the general public. The damage caused can only increase as a result of unawareness and irresponsible construction.

GS MITIGATION AND ENVIRONMENTAL ENHANCEMENT

It is important to remember that telluric forces are natural but constantly evolving influences; regrettably, we have little real understanding of them and perhaps even less concern as they stream into our environment, which is increasingly cluttered with vibrations emanating from power grids, electrical circuits, computers, mobile phones, and routers, amplified by metallic structures and furniture, all of which play havoc with our vibrational balance. Nature's creations are then left to struggle in their midst with very little help coming from the cosmic energy that in the past balanced the whole. And they will not go away just because we ignore them.

At this point, humans have had sufficient time to learn and become familiar with Nature's ways. Now while that is a very long time, surely much longer than the average history curriculum would have us believe, there is very little, if any, information that has worked its way into the practical stream of academic knowledge.

Any builder today with a few years of experience under their belt will be aware of the effects of certain terrains, yet probably be quite unable to explain the causes of geological phenomena. Subsidence is one of the more easily explained problems. The capillary action of subterranean water can only be determined on a microscale with radiesthesia; the excellent quality of the British hydrological survey maps around the world are most useful but on a much larger scale. Rising damp, poor drainage, and fungal issues can be happily, albeit expensively, resolved, and solutions are available for a limited number of other problems as well. For instance, GS can cause cracking in concrete or walls, often along the line of flow of the unseen influence, which frequently leads to insect invasion or infestation—they love the low-energy areas—and it is relatively easy to get rid of the pests using frequency-restoring systems.

However, we are, by and large, totally unaware of the forms and patterns assumed by energy in its movement. The exceptions, of course, being found in aerodynamics and certain disciplines of the physic sciences. These flows are inherent and essential to life, even if we do not appreciate that. Movement is the agent of change.

But there is every reason to believe that our forebears knew a lot about the energetic influences coming from the ground. We know this because they left proof of the remedies they found to deal with certain issues. Especially the ones that caused problems.

For instance, it is said that there is a stele in China on which the ruler of that particular time, Kuang Yu (2205–2197 BCE), had inscribed an edict that ordered the surveying of building sites to make sure those areas were not subject to underground water currents, so as to avoid those noxious influences but not to get rid of them. I think one can reasonably say that no ancient site over two thousand years old, anywhere in the world, has ever been found to be built on subterranean aquifers unless those currents were incorporated into the overall design, as is seen with the pagan precursors of the European cathedrals and churches of the Middle Ages, the menhirs and dolmens of earlier times, because by building them there one could remove that harmful influence. The medieval builders reputedly even tested their building sites first by putting sheep there and observing them to see if they avoided the areas, and it is well-known in agricultural communities that animals will not give birth over such spots. And today, the Chinese feng shui experts I have met and questioned have no solution to resolve geopathic stress for the simple reason that if you don't build over it in the first place, you don't have to remove it.

But here's the rub: we in the modern world have little option but to build where space is available, and there is not enough space to position all buildings out of harm's way. In Europe today, only the Germans are concerned about telluric influences, even requiring by law for a *Baubiolog* (radiesthesist) to check the site out first before authorizing the start of building.

It would appear, however, that ancient peoples such as the Irish, Celts, and Egyptians, realizing the potential utility of these subterranean forces, integrated them into their vestiges and practices. Early modern humans in all probability used, among other devices, large standing stones to transform the harmful effects of these influences to beneficial frequencies, or alternatively they steered clear of them and built their homes elsewhere. This type of geometric and natural "science" was probably transmitted at a later date by various orders— Templars, Cistercians, guild members—because the effects of these forces are demonstrably active even today despite the major modifications made in recent times to many European cathedrals and monuments built during the Middle Ages.

The ancients (Sumerians, or Babylonians rather, Chinese, Egyptians, Greeks [because they applied what they learned at school in Egypt] and Indians) knew not to build on zones that are affected by what is below ground and works its way up to wreak havoc on the surface. We know that fact by inference alone because, as previously mentioned, no ancient monument was ever built over such problem zones, which is an important fact to emphasize.

It is especially relevant to note that most, if not all, of the medieval cathedrals built in the twelfth and thirteenth centuries, were located above these powerful subterranean water passages. The architects seemed to have known that these sites were connected, so to speak, by these underground waterways, and the construction sites were chosen as a consequence of that awareness. Although that awareness antedated the cathedrals, which in many, if not most, cases were built on older, pagan sites. Once again, as we will infer from other examples later in the text, authorship of available knowledge is usurped and closeted by the powers that be.

Having realized the danger of these subterranean forces, and found ways to detect them, it is but a step to surprise a way to do something about them. And there can be no doubt that the ancients did, for they integrated those remedial measures into the vestiges, which we still have among us. It is from these that we are able to discover that they not only knew something that we no longer understand or that has been occulted, but they were knowingly applying those methods.

As previously mentioned the magnetic component of stone is closely related to water, largely due to water's distinct intelligence. When the two are combined to form a mutually supportive magnetic field, it seems that a lasting coherence can be created, resulting in a harmonious environment that can carry some distance as a function of the telluric influences that interfere, but only when reaching the surface, so that benefit can be conveyed several hundred yards or even several miles.

As also stated earlier, the most harmful telluric influence is the crossing of underground water, for while it may be a joy for the wyvern, it is very deleterious for man. Systematically, standing stones are to be found erected on the exit point of those currents, and it is no coincidence that the pillars of the medieval cathedrals are also erected on those points, as is very often the case with the standing stones, obelisks (when in their original positions), and *all* the authentic round towers of Ireland.

One can presume the reason for placing the stones, as we saw earlier with the experiment Schauberger made on the disappearing spring, was to encourage the spiraling pulsation and levity of water from the depths of the earth, drawing the wyvern up, so to speak. The advantage of course is that the effect also modifies the danger of those magnetic fields and in addition can create a space of intense harmony and peace for human, animal, and plant. What is more, in that zone the energetic influence is sometimes sufficient to change the frequency over much larger areas, including those used for farming.

Our forebears well understood that poor soil meant unhealthy plants and sick consumers. Crop-destroying bugs are Nature's scavengers, keeping the place tidy, so if you want health for one and all, the trick is to take care of the soil. If you don't have a ready supply of volcanic (paramagnetic) dust, use compost, correlate the weeds that bring the minerals up to the surface, develop mycorrhiza thanks to worm population in the earth, and of course, apply magnetic force, through the natural frequencies absorbed and retained in rock.

The use of stone by post-diluvian humans would appear to offer a key insight into the understanding of what the ancients knew about the magnetic force of stone. The palette of skills they used to offset or palliate the effects of telluric influences ranged from the use of large standing stones and obelisks to structured stone laying and geometric form, among other devices. It is still possible for us to perceive these, but it does require a certain flexibility of intellect and openness of mind.

This type of natural knowledge was probably closely guarded by the fraternities (the girls surely had better things to do, although they were probably members of the original gangs, too!), guilds, and groups. I would venture that such protective reasoning was not out of an interest in any form of social control but more a question of maintaining precision in the analysis and application. Alternatively, with a recent apocalypse in living memory, water and mud floods that destroyed most of the world some few thousand years ago, the survivors, unable to make heads or tails of the Egyptian hieroglyphs, decided written communication was not the way to go.

Naturally, at some stage such esoteric know-how is viewed as magic, and given that public opinion is easily corrupted, especially when vested religious interests are in the balance, there were probably some protective measures taken, such as not writing these notions down and committing them to memory of the select few, as was apparently the Druidic tradition.

It might be pure speculation, but the story of the Templars that follows in chapter 9 emphasizes this human tendency to strive for the upper hand when financial profit or control offers some benefit, or how things become perverted in the process as soon as greed is involved, and the greater benefit (for Nature at large) is soon forgotten.

Given the evidence the ancients left behind in the form of standing stones of various sorts, and there are at least sixty-five towers left in the world, despite so many having been destroyed or deliberately removed, it is astonishing that we have not figured out what is going on.

If we are really sincere in wanting to understand how we can fit harmoniously into Nature, we have every interest in observing and emulating Nature's ways rather than attempting to impose our ways on her. Such a way of thinking would presumably have been axiomatic in the minds of people who had survived disasters or were living in memory of such cataclysms that destroyed the dinosaurs, others that provoked the Ice Age, and so on.

The upheavals caused by the flood of the earth and the events that preceded it have, in all likelihood, left us with a very different living space from the one that existed beforehand. No matter if you believe that series of events or maintain that the earth is a more recent affair, the fact still remains, geologically and tellurically speaking, the place is currently complex, and we do not have a science worthy of the name to aid us in navigating it.

If there was a branch of learning that studied the effect of geology on the inhabitants of the troposphere, we would know that certain subterranean conditions are dangerous and ignored at our peril. There is no such branch of official learning today, but that does not mean it never existed, as indeed it did—and still does—but as radiesthesia.

THE COSMIC ASPECT

So far I have placed emphasis on the telluric interaction of this magnetic effect of stone—its interaction with the earth due to the water current below ground, and the evacuation of that energy to the skies where it seems to be "dealt with" by the incoming cosmic force. There are, however, other ways of considering the possible interaction of the forces involved.

The "antenna" approach, as proposed by Philip Callahan in his book *Ancient Mysteries, Modern Visions*, suggests that the round towers may have been designed and constructed to be used as huge resonant antennae for collecting and storing wavelengths of magnetic and electromagnetic energy coming from the skies. Based on his studies of the forms of insect antennae and their capacity to resonate to micrometer-long electromagnetic waves, Callahan suggests that the Irish round towers (and similarly shaped religious structures throughout the ancient world) were human-made antennae that collected subtle magnetic radiation from the sun and passed it on to monks meditating in the tower and plants growing around the tower's base.

He proposes the round towers were able to function in this way because of their form and also because of their materials of construction. Of the sixty-five towers, he states that twenty-five were built of limestone, thirteen of iron-rich red sandstone, and the rest of basalt, clay slate, or granite—all minerals that have paramagnetic properties and can thus act as magnetic antennae and energy conductors. Callahan further states that the mysterious fact that various towers were filled with rubble for portions of their interiors was not random but rather may have been a method of "tuning" the tower antenna so that it more precisely resonated with various cosmic frequencies.

Callahan also postulates that the geographical arrangement of the round towers throughout the Irish countryside mirrors the positions of the stars in the northern sky during the time of the winter solstice. Apart from the difficulty of earthbound humans to find any access to such a layout, there seems little, if any, practical use. Of all his theories concerning the raison d'être of the towers, this seems to be the least probable. As a radar technician responsible for the guidance of aircraft, he would most likely have been in possession of a copy of the Rude Star Finder and Identifier, published at the Hydrographic Office in Washington, DC in March 1942 under the authority of the Secretary of the Navy. This was a very useful device for a plane pilot/navigator or other person to plot their location from the position of the stars and planets in the sky. Those interested could compare a star finder map like the one on p. 112 with the map on p. 134 to get an idea of what Barrow proposed.

Quite what such a comparison adds to the discussion I fail to grasp, as it seems to be a sensational suggestion that the ancients felt a need to lay out a map of the stars on the ground in Ireland, or have I missed the point? Who would use such a map? How would they use it and for what purpose? The questions are multiple, the hypothetical answers are vague at best, and it seems to serve no practical purpose whatsoever.

As to his idea of the towers being antennae, once again a number of simple questions relegate the idea to the same basket of interesting, impossible-to-prove and, practically speaking, rather useless information when applied to the possible use of incoming frequencies in the absence of a method to benefit from such an arrangement.

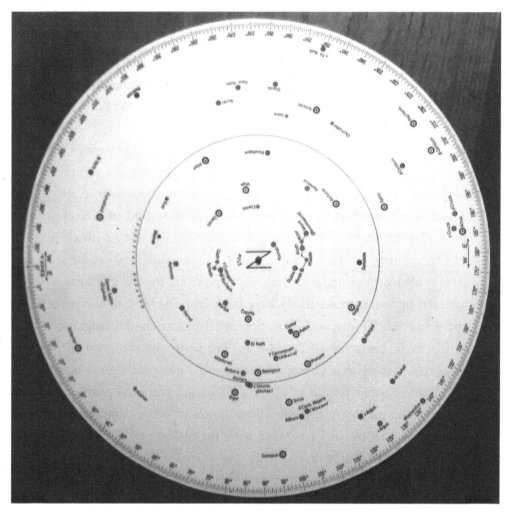

Rude Star Finder and Identifier.

The beauty of putting these ideas out there into the ether-space of people's minds is that there is a strong chance that somewhere down the line a small piece of the puzzle will be added to the whole.

My judgment of these ideas may appear harsh, particularly on someone who can no longer defend themselves, especially when and, what is more important, they rendered tremendous service to the subject. However, I stand by what I say as the situation today far exceeds any niceties we should politely extend to each other, for I am convinced that we are in serious need of remedies to resolve the sorry state Nature is in due to our abuse.

This takes us to the next stage of the investigation, but before we go there it might be a good idea to contemplate the possible motives for erecting the round towers in Ireland.

The erection of stones above ground in geometric patterns whether in groups or standing alone is generally easy for us to appreciate because we know from the agricultural standpoint how useful it is to know the time of year, from the astronomical viewpoint how instructive it is to predict and prepare, and from the ritualistic angle how beneficial for the local society. Imagine how those substantial benefits would be surpassed if the added value of well-being were to be included.

Consequently, assuming that there is a functional benefit in setting up a stone structure, if and when that benefit can be proven, subsequent to experimentation, to be the result of a specific means or method, there is a sound basis for presuming that those means are relevant in accomplishing that benefit and that its employment is what the builders intended by its application. That is the approach adopted in my research, which must then be considered, discussed, experimented, and explained—where possible. Practicality is of the essence—what works to our benefit without adversely impacting Nature at large. All other hypothetical and subjective considerations, as proposed by various proponents, are left for the reader to ponder.

Such a way of dealing with the subject matter immediately narrows the scope. By the term *practicality*, I imply necessity of the most simple kind: the need for survival of the individual, group, or community along with their well-being and continuity. It seems that without those components, the recipe for society soon disappears as people become disgruntled and move away, or at worst, die out.

The difference today is that we choose to ignore the fact that interfering with planetary and telluric frequencies is fraught with consequence. It is only when we respect Nature and remain within the bounds of reason that we can hope to maintain existence on an even keel. There is every indication that former mega-settlements were destroyed, perhaps as a result of man's folly, which encourages me to persist in this obscure research.

There are, however, several substantial shortcomings, as I see it, which could benefit from a human effort to restore our lost ability to commune with Nature in a more meaningful manner, and in so doing, achieve a greater harmony for all concerned. As you can intimate from the text above, a familiarity with magnetism is key. Another aspect that deserves greater focus to acquire a more complete view of the whole is water. The relationship between water and stone could be the subject of a substantial volume of literature but must be left for another day.

You will perhaps have appreciated that underground water is a potentially lethal component in the mix we are considering. It can, however, be turned to our benefit, and that is what the ancients appeared to be doing. The instances of underground water, especially the crossing of water currents, as a cause for serious problems are so numerous among radiesthesists, shamans, and dowsers, there is little room for coincidence in this respect.

Of interest and possibly of some importance is the magnetic layout of the stones used in the construction of the Great Pyramid of Giza. All the stones appear to be placed with the negative magnetic polarity at the top. And the whole edifice is built over the crossing of underground water. If the demonstration of this understanding and application of magnetism and water is still to be found in Egypt, there is a further avenue of possible explanation for that knowledge reaching Ireland with the arrival of the Egyptian Scota with her Greek husband, Gaythelos, and their company. But that is yet another rabbit hole!

7

The Three Principles

Presented in this chapter are three intriguing principles—Geographic, Polarity, and Cross—that may at first appear idiosyncratic but when developed further will help to support the theories presented here.

THE GEOGRAPHIC PRINCIPLE

Despite the major modifications made in recent times to many European cathedrals and monuments built during the Middle Ages, the effects of what I call the Geographic Principle are demonstrably active even today.

There are two main elements to the Geographic Principle: orientation and purpose. Orientation was a major principle in the siting of a building in the past. One obvious factor is the rising sun and our dependence on that body for our material existence, but it is far from the sole criterion. Chartres Cathedral is a notable exception, with its orientation toward the northeast, perhaps in alignment with the direction from which one of the main underground water streams flows.

However, before orienting a building—and even before finding the right place to build it—one needs to be quite decided on its fundamental purpose. Accepting the church-origin story put out over the centuries stops us from considering just what the ultimate purpose was in building and orienting a building in a particular place.

Can we seriously accept that forty or so years of construction work, at the vast expense of labor, money, and expertise, was simply to build a new style for a place of worship? That is what we are *told*, and most believe it without any further thought.

But what if the crossing point of the underground streams over which every cathedral was built in that period could be turned to advantage? One could then make as if the Roman church organization—and not the strategically placed stone cathedral itself—was the origin of the resultant well-being.

No matter if the pagan sites that apparently existed beforehand in those same localities used their own methods to remove the harmful effects of the water crossing—with well-placed round stones; the memory of the general public has never been too solid, especially when you distract their attention and provide them with something bigger and better to believe.

That makes for an extremely expensive marketing program, but what better way to demonstrate power and seize authority over the common mortal? Or what if, alternatively, the Templars paid for it all and were then conveniently "disappeared?"

I would maintain here that the emphasis on the objective—that was subsequently usurped by the Roman church in this instance—was for the well-being of the general public because once one has experienced such a sensation of comfort, even in part, one potentially gains considerable control. If one can offset the harmful effect of the most powerful telluric influence by conveying its force upward to the heavens, one can make the surrounding area on the surface of the earth welcoming and calm. That must have been, and still is, a promising accomplishment and a likely consideration to bear in mind as a reason for that objective.

But that is not the end to the matter. One also needs to know the method used to achieve that noble aim of well-being. In the case of the cathedrals as we can deduce today, that required a precise knowledge in order to calculate the "cubit" measurement of the locus.

The science that the Templars brought back with them was vast in potential, so there was probably some interesting reading stashed away in Solomon's Temple too. I would even venture that never since the construction of Cheops Pyramid has so much wisdom been potentially revealed. Are you aware that the British, in the twentieth century during the period between the world wars when they were "protecting" the Holy Land, were digging extensively in Jerusalem?

The Chartres Cathedral "number" is 73.8 centimeters, or according to Louis Charpentier in *Les Mystères de la Cathédrale de Chartres*, one hundred thousandth of the total distance longitude around the world on the same latitude where the cathedral is positioned. I cannot agree with this calculation, but playing with it allowed a more precise method to be revealed. This cubit of Chartres serves as the basic measurement that is combined, applied to, and proportioned throughout the construction of the building. This same geographic principle is applied and used for each and every cathedral built during that period when the expert craftsmen or companions were practicing their skills.

While there is no record of how these craft guilds came into being, it was most probably a form of schooling that developed. It is hard to imagine how else one could produce such a large number of skilled workers in so short a period of time and covering such new and varied trades as stonemasonry, carpentry, roofing, foundation laying, scaffolding, glasswork, coloring, rose-window stone lacework, flying buttress techniques, and so on.

The Geographic Principle, as I call it, is one of the keys to the mathematical computation of the structure, as well as, I firmly believe, the cause of the sensation of well-being developed when using that measurement. The reason for saying that is because this sensation of well-being can be easily reproduced when using that measurement specific to the geographic location of any position in question. This very same principle appears to have been used in just about every ancient building throughout the world, whether in Egypt, China, Peru, Mexico, Greece, or France.

That would imply that the knowledge required to implement this technique is old—much older than the Templar institution and Greece, and probably antedating Pharaonic Egypt. This leads to more questions, and even if we did find archaeological evidence at the bottom of the seas, we would probably still be none the wiser. So, let's stay in the domain of the possible, practical, and what we can physically work on.

It is of some consequence to learn that round river rocks, boulders, were discovered in the foundations of Chartres cathedral during renovation work in the 1960s. Regrettably no mention of magnetic polarity was recorded. Of course many of these European cathedrals and churches were built on older pagan sites, where remedial work had already been performed, which consequently benefitted the site thanks to the measures taken and not having to prospect for new sites.

THE POLARITY PRINCIPLE

It is now time to introduce the second principle, the Polarity Principle, which adds one more, and perhaps the most significant, ingredient to the brew. If a stone has a magnetic field of force, it is only a simple step to experiment with it as a function of its environment, makeup, and position to achieve the benefits that manifest when they are properly applied.

We are familiar with the benefit of modifying a plant's, animal's, or human's current state of well-being or health by performing the laying on of hands, magnetism, or Reiki (or chiromagnetism as I prefer to call it)—all examples of what man has always known and used throughout the ages. Could it not be reasonably assumed that this applied form of magnetism can be achieved through other sources and not just via the human medium?

Couldn't the same force be emanating from standing stones but with a much greater potency and with less variation than the field coming from an individual's hands? It would also be, in all probability, of a greater or lesser intensity as a function of its position in relation to the telluric influences over which it is placed and therefore provide a permanent output that could be felt by anyone or anything in its proximity.

This is perhaps the principle our ancestors understood and worked with when they erected the stones, whether obelisk, menhir, or round tower. For it seems that certain stones have a "healing" capacity that could possibly be due to this magnetic field providing a harmonious energy. As recounted above, such stones can be usefully employed to restore health in certain circumstances, but what is more intriguing is their ability to reestablish the energetic balance in the more general environment, which seems to be of equal importance in maintaining well-being, as it contributes to a more harmonious physical, mental, and emotional state. This is similar to the way that magnets, when applied to the skin over a broken bone, heal the bone faster than it would heal without them, while generally absorbing inflammation, more rapidly reducing pain, and making for a happier patient.

That is the aspect we are now going to look at a little closer. Would it not be too much of a coincidence for the negative magnetic polarity in stone structures to be systematically positioned toward the sky? If it was not a coincidence, it would indicate that such positioning was a purposeful architectural application of a recognized phenomenon. As previously mentioned, this same disposition of stones in building, negative above and positive below, is found in numerous structures around the world. So the theory is that ancient architects were aware of the positive and negative magnetic polarities in stone, and they knew that those properties, specifically in paramagnetic stones, were powerful stimulants—for agriculture, animals, and humans.

At this point, it may be helpful to take one step further back in time and look at another monumental vestige on our journey—the obelisk. Ancient monolithic obelisks distributed around the world all have something in common with the round towers: they share the same magnetic disposition. In other words the negative polarity is at the top and positive polarity is at the base; however, no one has noticed or commented on that fact.

We are led to believe that there are twenty-eight obelisks in all that came from Egypt, only seven of which are still in their country of origin. It is said that the obelisks were generally erected in pairs, as one finds in Karnak, but that is a later development and possibly not the original disposition. It may seem that placing two obelisks closely together would make them magnetically inimical, but if they were spaced in full understanding and knowledge of the forces at hand, there is a strong likelihood that a specific kind of magnetic field of force could be achieved. That would seem to be confirmed with the dolmens or standing stones, some of which appear to have been used for healing purposes, as tradition would hold in Ireland, and perhaps elsewhere.

Incidentally, the tallest obelisk in Egypt is to be found in Karnak, originally erected by Queen Hatshepsut, who was much maligned by her successor who removed many of the monuments erected during her long reign.

Most of the obelisks were inscribed with one or more columns of hieroglyphs on one or more sides (four columns on the four sides for Seti II in Karnak). There are, however, a few ancient obelisks with no inscriptions, including the Vatican obelisk that Caligula brought to Rome. (Rome today has thirteen Egyptian obelisks, which is more than those currently above ground in Egypt! What did they know?)

It is not practical to engrave stone when you are perched 60 feet up in the air, so it would be reasonable to assume that they were inscribed while they were still lying on the ground. The information in the inscriptions related and gave praise to the reign and title of the person concerned, although additional columns of hieroglyphs were added up to four hundred years later. But if they are not all inscribed to commemorate or extol the feats and works of the person commissioning the monument, why would builders go to the pains of quarrying such a large piece of masonry?

As is so often the case in Egypt, as elsewhere, there is more going on beyond the symbolism, grandeur, and egocentricity than meets the eye. Could there perhaps be a more grassroots pragmatic function for the obelisks? Maybe to evacuate the damaging effects of the crossing of underground water to thereby (1) create a space of well-being for whoever spent time in the neighborhood, (2) broadcast and release vibrations from below ground with an antenna, or (3) draw in energy from the atmosphere at a certain height?

Creating space for well-being is achieved when the attracting force of the stone obelisk pulls the harmful-to-human magnetic field caused by the combined frequencies of the water crossing upward and away into the atmosphere. This same principle is applied when building a pillar in a cathedral or church. Broadcasting vibration and drawing energy from the atmosphere would be achieved by a concentration of negative ions surrounding the negative polarity of the obelisk. An obelisk as well as a round tower can achieve all of these functions.

As is to be expected, there is controversy surrounding the origin of the Egyptian word that might define an obelisk. We have the term *tejen*, which in Egyptian signifies protection or defense, and *tekhenu*, meaning "to pierce." This last is reminiscent of the Thai *chofah* (Sanskrit: *cho*, meaning "to cut, pierce;" Thai: *fah*, meaning "sky"), that part of a Buddhist wat (temple) or palace roof architecture where a decorative Garuda-type addition "pierces the sky."

But what would these obelisks be protection against? If my proposition is acceptable, the answer seems obvious: the harmful effects of an underground water crossing. There does not seem to be any other logical explanation, but as always, if there are differing theories, I'm all ears.

The problem, and it is one of consequence, is that there is only one obelisk in Egypt—Sesostris—that is still in its original location, or so we are led to believe. When asked whether that is actually the case and if the obelisk is over the crossing of underground streams, a pendulum answers affirmative to both. Sesostris is the oldest known obelisk, according to the pharaoh.se website, and there is a crossing of subterranean currents below it, even three thousand-odd years after its erection. Despite the probability of the Nile changing course since the time when the pyramids were built in Giza, it would appear, according

Sesostris obeslisk at Heliopolis.

to questions with the pendulum that the Sesotris crossing existed when the obelisk was erected.

Theories abound concerning the possible purpose of obelisks. Hardly in agreement, suggestions run the gamut from being revered as sacred objects, to being erected as shows of luxury, to symbolizing shafts of sunlight, to commemorating events, to demonstrating power, or even serving no purpose at all! Such wide variation serves to demonstrate not only our amazing imaginative ability, but also a more concerning issue—our serious lack of gnosis.

At the very least, it's hard to imagine that obelisks *wouldn't* represent a special power. If, as is the case, other stones erected around the world reveal the same negative magnetic polarity up and positive down, why is no attention given to this fact, and why is the emphasis placed on nonsense instead?

If we are attentive and know where to look, there is evidence throughout the world that former civilizations were not only aware of these issues but took active measures to ensure that these telluric influences were attenuated when overly aggressive or avoided altogether. What is even more pertinent is that it was often possible to transform the harmful effects and enhance the benefits.

In Ireland alone there are between forty-five and fifty thousand stone circles, according to the excellent *Sun Circles of Ireland*, written by Jack Roberts in 2013. It is not just the stones in the circles that demonstrate this magnetic polarity, the towers do as well. The difference being that each and every stone in a tower is positioned in such a way that the Polarity Principle is achieved, thereby multiplying the field of force in consequence.

Numerous examples of buildings featuring the Polarity Principle are to be found around the world. To any reader interested in studying them further, I would suggest starting by casting a careful eye on the Lion Gate in Mycenae Greece, the Inca palace in Machu Picchu, and the Tiopunco portal of Sacsahuamán in Cusco, Peru. What is more, these monumental structures are magnificent demonstrations of Cyclopean masonry, with their carefully devised and implemented interlocking mechanisms that are able to resist earth movements and aggression of all sorts.

THE CROSS PRINCIPLE

Nearly ninety-seven percent of the earth's usable freshwater is apparently underground, and seventy-three percent of the surface of the earth is covered by water. That leaves a lot of room for investigating and researching the nature and properties of water, but it also gives us an indication of the importance water plays in our biotope.

As explained in some detail in chapter 6 and elsewhere, one of the most harmful conditions found for human and animal life is the crossing of underground water. The damage caused by this phenomenon was known and considered so detrimental that in the Chinese tradition, one does not build a house over such a geological event. That presupposes that it was a known danger to be avoided because problems had arisen at some stage in the past and that someone knew what the origin of the trouble was and, most importantly, how to locate it.

Underground streams are made up of water that flows through

any underground passage, fracture, or fissure. As it does so it produces its own electromagnetic field often high into microwave frequency, as well as generating flows of ions, generally positive ones. This field will fluctuate depending on what is dissolved in it (e.g., sewage, grave-yard or dumping ground sediment, fertilizers, and insecticides), how fast it is flowing, and whether it is interacting with any other type of earth energy. Interference with the earth's natural energy field is particularly marked by the flow of these underground streams, especially where two watercourses or other types of energy lines cross. In a league table of radiation associated with biological damage, the out-side line of a subterranean water course would be at the very top.

These are the natural but evolving entities that we are dealing with, albeit with little, if any, real understanding and regrettably, with a widespread lack of concern, as they stream into our environment that is already cluttered by vibrations emanating from the many technologies of today, all of which play havoc with our vibrational balance. And once again, there stands the individual human being struggling in their midst with very little help coming from the cosmic energy that could balance the whole, if we were in an ideal setting.

When you realize that the round towers, as well as many other old structures are built on these hazardous-for-human zones, you cannot help but wonder what made the builders reckon they had some kind of remedy that could mitigate the danger, and additionally, improve the situation. This would indicate that they not only knew of the problem but that they also had the expertise to transform that energy into some-thing beneficial, something *healing*—drawing up a source of energy, redirecting it to where it is no longer causing liability, and then turning it into an asset for a larger community.

8

Key Elements of the Round Towers

An island in the ocean over against Gaul, to the north,
and not inferior in size to Siciliy, the soil of which is so
fruitful that they mow there twice in the year.

DIODORUS SICULUS, 60 BCE

Now that we have had a closer look at what makes the essence of a round tower, it might be helpful to run through the technique of how to verify those factors are present. There are a number of ways to do this, requiring either one's physical presence at the site of a tower, or remotely when a lot of that work is very much facilitated thanks to Google. By calling up a round tower onto Google Earth, one can position oneself immediately above the tower, and so gain the exact bird's eye location required for an instant answer to questions with the pendulum: "Is this tower located over the crossing of underground streams?" or "What, if any, form of telluric force is present here?" The process might be somewhat more laborious if you are physically beside the tower, as when asking the same question you must place your questioning self in the axis of the tower, so to speak.

It is up to you whether you want to discover the depth and direction of the water currents, which is always useful data if you would like to find out if there some form of connection between towers, waterways, megalithic sites, fairy forts, and so on.

If Google allows you to view the tower from street level (rarely the case), then you can move to the next phase of checking the polarity alignment of each stone from the image that appears on your screen. If you are present in person, using your free hand as a pointer, ask where is the negative magnetic polarity. This can be a lengthy procedure and shortened considerably by asking if any of the stones in your view/in this tower are positioned with the positive magnetic polarity facing up. There will potentially, of course, be some, because of past repair work, but surprisingly few, and I wonder if the magnetic alignment is not "forced" over time to conform to the general pattern. That would be an experiment to try, and quite easy as some of the towers are accessible and the inside walls damaged, so allowing such sacrilege.

The geographic principle is manifest in a far more subtle manner. By taking a reading of the overall energetic score in the immediate surroundings, you will find if there is a difference indicative of the effect developed by the combined crossing and polarity principles. There is a good chance that a more sensitive person will feel the influence of water far more than other telluric forces, and that is generally the case when in the precincts of a round tower.

Thales of Miletus (circa 614 BCE) was on to something when he said that water is the only true element, as all the others are derived from it. Another Greek, Plato, states in his *Timaeus and Critias* that the people of Atlantis regulated water using its temperature, as water needs a balanced, organized, and regulated temperature system to develop. Until such time as we have a better if not complete understanding of water, we are doomed to a stopgap strategy regarding water management, and probably as a consequence, all aspects of agriculture. If we want a more complete picture of water, we need to take a look at what happens to it when it is underground, where it seems to spend a substantial share its life cycle. For it is from there that I believe the round towers are drawing their influence.

The fall in groundwater, or the water table, is perhaps the main reason for failing crops, as they lack the necessary mineral nutrients that are brought to the surface with the flow of water from below. Rainwater only moistens the earth; there are practically no nutrients to be absorbed by the soil when it rains.

The natural cycle involving the build-up and decomposition of vegetation includes a layering of the substances in that vegetation, similar to the process of composting. Because there are substantially different temperature and pressure constraints found underground, those substances are changed into carbon compounds with the help of water and local conditions. It would be more than probable that the water is also decomposed in this process, resulting in the production of more gases that release further carbon compounds, such as carbon dioxide and/or carbonic acid, which then start their journey up to the surface through the layers of the vegetation above, causing further movement and transformation in the chemical compounds in their passage. Given that the air mixture down there in the bowels of the earth is quite different from the conditions found in our biotope and is perhaps even absent, totally new conditions and possibly compounds are developed, leaving us pretty much in the dark as to the movements and transformations underway.

We can only deduce what that alchemy is from the end products that reach us, which will then be further modified thanks to the help of the kinetic force from the magnetically aligned stones in the tower drawing them upward. It seems clear that the enclosed environment below ground is the source of new forms of vegetation and substances such as ore and rock, adding to the potentially magnetic mix with its influence on water and the surrounding substances. This is all in line with Nature's governance, ruling temperature, and pressure and their exchanges, which give a coherent and sequential picture that, if we observe more closely, might help us get a better handle on the processes involved. Those intricate procedures occurring at varying depths and subject to changes in temperature and pressure and to internal movements all form natural laws that we can deduce if we are attentive. There is a strong chance that in past times the contemporaries of the builders of the towers might well have understood them.

What exactly is needed for growth? Where does the right balance lie? Obviously, Nature holds the key, so every effort must be made to respect what we observe and profit from in this cycle. For it is thanks to functional events in the correct sequence and conjunction that refinement of the substances in question is accomplished, in turn leading to further avatars (in the Sanskrit sense of incarnation) in due form and time. This all happens very much thanks to the special movement (spiraling perhaps) of water, as it oscillates between kinetic and potential energy, generating the electromagnetic current and its accompanying magnetic phenomenon, which of course are not static and evolve in their progress, interacting as they move on their merry way.

Nowhere else can one find such a satisfactory explanation that fundamentally expresses the mechanics and probable physics that were appreciated by those beings who built the round towers. There can be no coincidence for there are a number of details systematically found in common among all authentic towers that would logically preclude any question of haphazard chance in the deliberate and consistent application:

- All round towers in Ireland are located exactly above the crossing point of underground waterways.
- Every single stone is laid with the negative magnetic polarity facing upward.
- Every tower is made with two walls, one external and one internal, reinforcing the draw capacity.

■ ■ ■

Why would you pay for a mason to dress every stone of a hundred foot tall building, not with just one wall but very often two, and then stack the stones in a very specific order? Dressing stone, for those not familiar with a mason's work, is a tedious task at the best of times, complicated by the hardness of the stone. It involves cutting the stone to the curve that the roundness of the final structure determines and requires a certain skill if one is to respect the rules of the art. The towers taper as they rise, so the diameter of the walls recedes as one gets higher, requiring the stones to be increasingly rounder on their external face. And this was often accomplished both on the inside and outside walls. That represents an awful lot of work and dedication. We should not forget either that a mason's work was generally performed outside, exposed to the elements at all times, even if a rudimentary shelter was erected to provide some comfort. Masonry was a harsh and taxing trade that most likely left most stonecutters in poor health at a relatively early age.

A very simple instance is never mentioned when dealing with ancient construction work—the cost. Not that the expense is of any great consequence to the whole question, because we do not know what the benefits were most of the time, and we never shall in all probability, so it is not a relevant consideration until one ponders the reason for going to such pains. One might reasonably assume that the benefit was worth all the effort. When that effort is reproduced elsewhere as in the case of one hundred or so towers that once existed in Ireland, there must have been a substantial advantage for all that work, otherwise there would be little point in doing it.

If an open mind could be brought to bear on the points raised here, I believe it would be a small but sure step toward learning more about the peculiar and unique energy generated by these towers. Perhaps that might lead to a better understanding of their objective, which as I have suggested, had something to do with well-being of the local community and the environment.

It is difficult, sometimes impossible, nowadays to access the towers, for reasons of security and inaccessibility for a host of motives. Even if one were able to do so, measuring sealed doorways, climbing up to heights without special equipment to verify orientation and measurements, to say nothing of the possible damage one might cause, would serve no purpose and would not make us any the wiser.

However, if we take a leaf out of Philip Callahan's book *Ancient Mysteries, Modern Visions* where he mentions the benefits of strong paramagnetic fields on agriculture, supported by ideas expressed by Albert Roy Davis and Walter C. Rawls in *Magnetism and Its Effects on the Living System* and Douglas Dunlop and Barbara Schmidt in *Biological Effects of Magnetic Fields,* we can start to comprehend the possible aim the ancients had in mind when they stacked a pile of stones, 100-odd feet high, in the same magnetic alignment. You are going to obtain something very powerful thanks to the intense magnetic field of force created. But what is that something?

I would venture, on the strength of what I know from my experiments with stone and plant and animal life (including humans), investigation of old traditions from around the world, and an understanding of life sciences, that such a strong magnetic force is a way to boost the movement of the life force, so enhancing—perhaps even encouraging, as it is alive—the component elements of that mysterious essence.

It would appear to be an accepted physical fact that two electromagnetic fields imposed on each other at 90 degrees is a classic scalar formation that results in a carrier wave. (Incidentally, for the radiesthesists reading this, the Universal Pendulum indicates that the frequency coming from a round tower is white in the magnetic phase.) And that carrier wave, an analogy for which would be the electric phase that carries the current to the light bulb, coming from the energies generated by the electromagnetic fields of the water currents below combining as they rise to the surface of the earth is retained within the stone structure, which seemingly directs that energy in the alignment of the negative magnetic polarity until such time that it is released through the doorway and windows of the tower.

When combining the magnetic field of force from the polarity-aligned stones with the carrier wave coming from the crossing of underground streams, the resulting effect is what would appear to be the high energy that the ancients recognized as what Nature needs. Given that a pendulum can indicate the direction of a flow of water, it is normal when examining a tower to ask it which directions the flows of crossing water are going. It is from there that one can deduce that the direction the doorway is oriented is determined as a function of the flow downstream of the main/larger water current.

In his 1995 book *Paramagnetism*, Philip Callahan relates that the fisherman ferrying him out to the Devenish Island tower remarked that the grass was finer on the island than on the mainland. And when Callahan visited the Glendalough tower he recorded a 20-millivolt reading at the doorway with his specially designed device mentioned earlier. Quite what the significance of such a score is he unfortunately does not say. Callahan designed his Photonic Ionic Cloth Radio Amplifier to measure frequencies. He was able to detect wavelengths in the ELF (Extremely Low Frequency) and VLF (Very Low Frequency) diapason detecting several segments of the Hertzian ranges, from the brain wave region 8-33 Hz, the electrical anesthesia region 600-4000 Hz and the lightning segment, 25000-50000 Hz. What his readings signified he didn't venture to say, but he was clearly impressed by the differences found between ground level, where he started, and the level of the tower doorway. In my work of resolving geopathic stress, using a pendulum and chart to measure the difference between before and after, the maximum reading I find once the GS is removed, is +20, and given when using the same scale the maximum human electric reading is +60, there is good reason to believe this is in the mV parameter.

Since we share no reference points with what the ancients were doing, it is no wonder that our science cannot correspond to what we might be looking at here. There is a chasm that cannot be bridged for the simple reason that we are lacking all means to relate our modern world to what went down all those years ago. This is not a comfortable place for our mindset, which is encouraged to be in control of our environment. The only solution, as I see it, might be to expand our awareness to look at factors that do not generally appear on our radar.

Mention above was made to the idea of benefit for the community. If we can make an attempt at reading the minds of those who lived when the towers came into existence, there is a strong chance that they, like us all today, were primarily concerned about physical survival and about satisfying the instinctual needs of all mammals.

There is an ever-present factor that should not be overlooked here, especially in the Irish environment where such care was paid to traditional mores and customs. If the land surrounding a tower is reputed to be more productive and nourishing agriculturally speaking, as we learned from the incident of Callahan being ferried over to Devenish, that merits investigation because it is possibly one of the main reasons for their creation. When viewed from the purely practical standpoint, one may ask, "What does a tower do?" In order to learn more, we must consider the likely context of the time when they were constructed.

Ireland is an immensely rich country agriculturally speaking, as was even apparent in the days of Diodorus Siculus, the ancient Greek historian (see the quote at the beginning of the chapter). Were the weather patterns so different then that two different harvests of grass were possible, or was there something else going on?

There was no doubt a time when the country was covered by more forest than it is now, which meant that there was less open land to grow crops and an increasing need to produce on the available space. This is all very hypothetical, but the location and density of the round towers in open spaces for the most part, propitious for the growing of crops, would indicate such an argument has some validity, especially when you consider the direction of the doorway, which is where the maximum magnetic force is directed, laying down a magnetic field that carries and can be measured from a distance of almost 1,200 feet.

There is an additional element to the puzzle here. What if the towers connected to each other not only via the underground waterways, but on the surface? The density of their presence—and that is especially so in the central areas of the country on land conducive to growing. As you can see on the Round Tower map of Ireland, the majority of towers are to be found in the center of the country, which could perhaps be explained by some form of grid created to extend the benefit. When setting up a group of sandpaper towers in the garden, this effect seems to project the enhanced growing capacity beyond the perimeter achieved with a single tower.

The energetic output from a tower as provided in the Appendix gives a reading for some of the standing towers. These readings, the RR (radiesthetic reading), are probably much weaker now than when the towers were in operation for the very simple reason that they have all been revamped over time—or they have been destroyed or damaged or are missing their caps, either original or according to design, whatever that might have been.

That charge—magnetic or otherwise—can be demonstrated by experimentation and is the result of placing a structure over an energy source. For example, if you take a conical form with a hole in the apex and place it over an energy source, what exits the hole is a concentrated form of that energy, much like the energetic charge coming from the top of a pyramid. This is what is happening with the round towers built in the manner explained above, positioned over the crossing of underground water, and then directed over the selected area of country thanks to the "doorway" and the windows. This disposition has a double purpose: to enhance plant growth and to provide well-being by removing the harmful influence of the crossing of underground water and disseminating over the countryside a nourishing form of energy that might not have been there previously, at least not to that extent. In any case the danger is so threatening in its effect that Chinese *feng shui* and Indian *vastu* both warn against it.

Placement of the round towers in Ireland. Adapted from George Lennox Barrow's
Round Towers of Ireland, *Academy Press, 1979.*

DETAILS ON SELECT TOWERS

All the towers listed below were built over the crossing of underground water. So it would seem pertinent to ask what the energetic score of the field of force at that locality is, using radiesthesia. This measurement is what I refer to as the radiesthetic reading The extent of that force depends on a variety of factors, which we may be able to assess or not, including the time of day, other inputs and obstacles (e.g., power lines, phone masts), temperature and atmospheric pressure, and so on.

Some towers have been eliminated from the generally accepted list as they do not fit the principles found in authentic towers as expressed in this book—namely, that the stones are systematically laid over the crossing of underground water with the negative magnetic polarity facing upward. These include Egilsay in the Orkneys, and the towers at Ferns, on Ram's Island, and Peel on St. Patrick's Isle off the Isle of Man.

The information below was acquired by using my dowsing capacity and the Universal Pendulum with six hemispheres, as devised by the two French radiesthesists, Andre de Belizal and Leon Chaumery, a chart of my manufacture, and my limited understanding. (Grateful thanks, too, to K. Schorr, the creator of the database formerly hosted on the now-defunct website roundtowers.org from which some of the information is drawn.)

For each of the fifty-three towers listed below, the following information has been provided when available:

Geographic coordinates
Elevation above sea level
Radiesthetic reading
Diameter of the tower at the base
Circumference of tower
Height of tower
Additional details such as tower material, number and direction of
    openings, and height and direction of door.

COUNTY ANTRIM

Antrim Round Tower

Geographic coordinates: 54°43'26" N, 6°12'31" W
Elevation: 114.8 ft
Radiesthetic reading: 40
Diameter: 15.7 ft
Circumference: 49.3 ft
Height: 91.9 ft

Additional details: The northeast-facing doorway is approximately 7.7 feet above the top offset. The lintel is composed of a large slab of granite, as are the four side stones of the doorway, in contrast to the rough local basalt rubblework of the rest of the tower. The windows are all linteled with the same rough stone as the rest of the tower, and most are fitted with simple wood frames and glass. The top story windows face the traditional compass points (almost) and are smaller than the other windows in the tower. The other windows, in ascending order, face east-northeast, south, west, and again south. Eight windows in all. The surroundings were cleared in 1800.

Other items of interest: A large boulder with two sizable bullauns lies approximately 20 feet from the tower, slightly to the left front of the doorway. The name Antrim was originally Aontreibh, Irish for "single ridge."

Armoy Round Tower

Geographic coordinates: 55°08'05" N, 6°18'37" W
Elevation: 105 ft
Radiesthetic reading: 45
Diameter: 15.1 ft
Circumference: 47.5 ft
Height: 35.4 ft above ground, 6.5 ft underground

Additional details: Excavated in 1843, the tower is composed of laminated schistose sandstone (mountain freeze). The south-facing doorway is just 5.2 feet above ground level. The very narrow doorway arch is cut from a single massive lintel with decorative raised molding. There are no surviving windows.

Ram's Island

Geographic coordinates: 54°35'06" N, 6°18'19" W
Elevation: 72.2 ft
Radiesthetic reading: -2
Diameter: 13.6 ft
Circumference: 39.7 ft
Height: 42 ft

Additional details: Ram's Island not technically considered a round tower. The tower is capless and composed of rough masonry. The door facing southwest is .49 feet above ground and was filled in around 1940. The original southeast doorway was around 15 feet high.

COUNTY CARLOW

St. Mullins Round Tower

Geographic coordinates: 52°29'20" N, 6°55'38" W
Elevation: 190 ft
Radiesthetic reading: 20
Diameter: 16.78 ft
Circumference: 52.8 ft
Height: 3.3 ft

Additional details: Little more than a foundation, this round tower stump has no doors or windows. It is evenly coursed of granite blocks dressed to curve both internally and externally with rubble between the walls.

COUNTY CAVAN

Drumlane Round Tower

Geographic coordinates: 54°03'29" N, 7°28'44" W
Elevation: 187 ft
Radiesthetic reading: 45
Diameter: 16.7 ft
Circumference: 52.5 ft
Height: 38.1 ft

Additional details: The tower is incomplete, has a level top, and is composed of pale brown limestone. For the first 22 feet, the stones are well cut and closely fitted; the remainder are course and rough. The door, which faces southeast, is at 9 feet. There is one window directly above the door at 23 feet high. The tower was excavated in 1844.

COUNTY CLARE

Drumcliff Round Tower

Geographic coordinates: 52°52'05" N, 8°59'50" W
Elevation: 36.1 ft
Radiesthetic reading: 53
Diameter: 16 ft
Circumference: 50.3 ft
Height: 36 ft

Additional details: Limestone. On the lower and more level south side, the tower reaches a mere 7.9 feet. There is no door, and the windows no longer exist, although a report in 1808 mentions the door at 19.7 feet high and three windows, one facing west, another east. While the doors and windows no longer exist, the large breach offers a fine view into the construction of this tower. Most of the courses of stone are quite even and the interior stone is dressed quite smoothly. The traditional "sandwiching" of an outer and inner wall filled with rubble is quite evident, as are the floor corbels on at least one floor.

Dysert O'Dea Round Tower

Geographic coordinates: 52°54'34" N, 9°04'06" W
Elevation: 78.7 ft
Radiesthetic reading: 38
Diameter: 19.3 ft
Circumference: 60.8 ft
Height: 47.9 ft

Additional details: The tower is composed of well-coursed large limestone blocks, dressed to the curve. At the base, the external circumference is 60.7 feet, one of the widest of the recorded towers. The wide doorway is facing east at 14.6 feet high and is unembellished and arched with six blocks running right through, with the keystone slightly protruding. The left jamb is comprised of three stones, and the right has four stones.

Iniscealtra Round Tower (Holy Island)

Geographic coordinates: 52°54'56" N, 8°26'53" W
Elevation: 131.2 ft
Radiesthetic reading: 50
Diameter: 15 ft
Circumference: 47.2 ft
Height: 73.2 ft

Additional details: The east-northeast-facing arched doorway is 10 feet above the ground. The three stones in its arch run the entire depth of the doorway. The north-facing angle-headed window at about the second floor level, 26.2 feet high, is especially well cut with finely dressed stone. There are three other linteled, square-headed windows facing east-northeast, south-southwest, and northwest. The tower was excavated in 1976.

Killinaboy Round Tower

Geographic coordinates: 52°58'13 N, 9°05'05 W
Elevation: 131.2 ft
Radiesthetic reading: 40
Diameter: 16.7 ft
Circumference: 52.3 ft
Height: ~11 ft

Additional details: This is a featureless stump of roughly coursed limestone.

Scattery Round Tower

Geographic coordinates: 52°36'51" N, 9°31'00" W
Elevation: 29.5 ft
Radiesthetic reading: 47
Diameter: 16.6 ft
Circumference: 52.3 ft
Height: 85.3 ft

Additional details: The tower has a unique doorway at ground level, corbelled in three stones, and facing east-southeast. In ascending order, there are five lower windows facing north, south, southeast, west-northwest, and another facing north. There are four slightly larger windows at the top facing northeast, southeast, southwest, and northwest. The first window is immediately above the doorway. The tower was repaired in 1855.

COUNTY CORK

Cloyne Round Tower

Geographic coordinates: 51°51'43" N, 8°07'13" W
Elevation: 82 ft

Radiesthetic reading: 50
Diameter: 17 ft
Circumference: 53.3 ft
Height: 100 ft

Additional details: The tower is composed of purplish sandstone, dressed to the curve. The door is 11.2 feet high, and there are nine windows. The tower was excavated in 1841. Vaulting was repaired and replaced with battlements in 1813 or so. The bell in the tower dates from 1857.

Kinneigh Round Tower
Geographic coordinates: 51°45'51" N, 8°58'30" W
Elevation: 337.9 ft
Radiesthetic reading: 60
Diameter: 17 ft
Circumference: 63.1 ft
Height: 67.3 ft

Additional details: The tower has a unique hexagonal base and is composed of local slatestone, dressed to the curve. The door is 11.2 feet high, and there are a total of nine windows. The four top windows face south, north-northeast, west, and east-southeast.

COUNTY DONEGAL

Tory Island Round Tower
Geographic coordinates: 55°15'51" N, 8°13'55" W
Elevation: 13.1 ft
Radiesthetic reading: 50
Diameter: 17 ft
Circumference: 51.5 ft
Height: 42 ft

Additional details: The tower is composed of undressed granite beach boulders. The door is 8.6 feet high and faces east-southeast. Four windows face east-northeast, south-southeast, west-southwest and north-northwest.

COUNTY DOWN

Drumbo Round Tower

Geographic coordinates: 54°31'00" N, 5°57'35" W
Elevation: 374 ft
Radiesthetic reading: 45
Diameter: 16.4 ft
Circumference: 51.6 ft
Height: 33.6 ft

Additional details: The tower is composed of spalled rubble of local clay slate. The door is 4.9 ft high and is on the east. There is one window on the north.

Maghera Round Tower

Geographic coordinates: 54°14'17" N, 5°53'50" W
Elevation: 16.4 ft
Radiesthetic reading: 42
Diameter: 15.9 ft
Circumference: 50 ft
Height: 33.6 ft

Additional details: This is a stump composed of mixed stone, partly granite. The door is 4.9 feet high on the east, and there is one window on the north. The Office of Public Works (OPW) made repairs and did some rebuilding in 1877/1878.

Nendrum Round Tower

Geographic coordinates: 54°29'54" N, 5°38'51" W
Elevation: 52.5 ft

Radiesthetic reading: 30
Diameter: 15.9 ft
Circumference: 43.6 ft
Height: 14.4 ft

Additional details: This is a stump of mixed stone, partly granite.
Completely excavated from 1922–1924.

COUNTY DUBLIN

Clondalkin Round Tower

Geographic coordinates: 54°29'54" N, 5°38'51" W
Elevation: 52.5 ft
Radiesthetic reading: 30
Diameter: 13.2 ft
Circumference: 41.6 ft
Height: 79.6 ft

Additional details: This is a complete tower of poorly coursed (not
dressed) limestone with some granite that has a bulging base. The door
is 12.8 feet high and faces east. The four windows face as follows: south
at 26.8 ft, west at 41.2 ft, north at 70.5 ft, and east-south-west also at
70.5 ft.

Lusk Round Tower

Geographic coordinates: 53°31'34" N, 6°10'01" W
Elevation: 88.6 ft
Radiesthetic reading: 40
Diameter: 16.6 ft
Circumference: 52.5 ft
Height: 87.1 ft

Additional details: Limestone and sandstone door. The tower is complete to the cornice with a cap at 108.2 feet. The door is at 3 feet facing

east-southeast, and the ground was probably raised 8.2 feet. The tower was substantially modified when a belfry was built right up against the tower around 1500. There are six windows plus two blocked: northeast, east-southeast, west, north-northeast, southeast.

Swords Round Tower

Geographic coordinates: 53°27'27" N, 6°13'28" W
Elevation: 101.7 ft
Radiesthetic reading: 50
Diameter: 16.4 ft
Circumference: 52.5 ft
Height: 85.3 ft

Additional details: Limestone. The cap is deformed. The linteled doorway, just 2.3 feet above the present ground level, faces east and is fitted with a modern metal-grilled door. There is a large east-facing window with a slightly jutting sill stone at the second story level. The third, fourth, and fifth floors each have a small square-headed windows to the north, south, and west, respectively. At the top story are four large crumbling windows with flat arches (some with red brick, indicating repair work) roughly facing the cardinal compass points. This has been heavily repaired.

COUNTY FERMANAGH

Devenish Round Tower

Geographic coordinates: 54°22'14" N, 7°39'21" W
Elevation: 164 ft
Radiesthetic reading: 65
Diameter: 15.8 ft
Circumference: 49.7 ft
Height: 82 ft

Additional details: The tower is in excellent condition, well dressed and finely coursed. It is composed of sandstone with some limestone internally. The northeast-facing doorway is approximately 8.9 feet above the ground. The arch is composed of three dressed stones running the entire depth of the doorway. A plain raised molding frames the entire door, including the door sill. The single angle-headed window is slightly to the right and above the doorway at the second story level. All other windows are linteled with the third floor window facing north-northwest and the fourth floor window facing south-southeast. The fifth floor has the traditional four windows slightly off the cardinal compass points but with a human head carved over each. The tower was repaired in 1835 and 1971.

COUNTY GALWAY

Kilmacduagh Round Tower

Geographic coordinates: 53°02'59" N, 8°53'18" W
Elevation: 59 ft
Radiesthetic reading: 65
Diameter: 18.6 ft
Circumference: 59 ft
Height: 111.5 ft

Additional details: The tower is composed of limestone and is the tallest round tower in existence. The doorway, facing east-northeast is also extraordinary in that it is almost 26.2 feet above ground level. There are eleven angle-headed windows, the largest number of windows of any existing round tower. The extreme height of the doorway, the number of bell-story windows, and the significant lean to the southwest all make this a quite unique tower. The cap on the drum is also unusual in that it doesn't sit atop a cornice but overhangs the drum instead. The walls are over six feet thick at the base. The tower was excavated and repaired in 1878–1879 and 1971.

COUNTY KERRY

Rattoo Round Tower

Geographic coordinates: 52 26'23" N, 9°39'00" W
Elevation: 49.2 ft
Radiesthetic reading: 70
Diameter: 15.1 ft
Circumference: 47.5 ft
Height: 88.6 ft

Additional details: The arched molded doorway is 9.3 feet high and faces southeast. A single window in the drum of the tower on the fourth story directly above the doorway was carved from a single stone. The four large top-story windows are all angle-headed and roughly face the cardinal compass points. A curvilinear design runs along the top of the arch, resting on the wide raised molding of the arch itself and terminating in scrolls to either side. The tower is made of a variety of large purple, brown, gray, and red sandstone blocks, regularly spaced with smaller stones at intervals. In 1880–1881, the OPW reset part of the conical cap. A sheela na gig is carved on the top left-hand corner of the north window on the inside.

COUNTY KILDARE

Castledermot Round Tower

Geographic coordinates: 52°54'37" N, 6°50'05" W
Elevation: 262.6 ft
Radiesthetic reading: 20
Diameter: 15.6 ft
Circumference: 49 ft
Height: 65.6 ft

Additional details: The tower is composed of granite boulders with limestone spalls. The original cap was replaced with battlements. The doorway is at ground level. This is the only tower with a window in the same story as the doorway. There are two small flat-headed windows in the body of the tower: one to the southeast and another to the south. In the top story there are four windows, each facing a cardinal compass point.

Kilcullen Round Tower

Geographic coordinates: 53°06'27" N, 6°45'39" W
Elevation: 515.1 ft
Radiesthetic reading: 35
Diameter: 16.4 ft
Circumference: 47.6 ft
Height: 32.8 ft

Additional details: The tower is made of local slatestone except for most of the doorway, which is granite. The doorway is 5.9 feet above ground level. There is one rectangular window, which faces southwest.

Kildare Round Tower

Geographic coordinates: 53°27'27" N, 6°13'28" W
Elevation: 101.7 ft
Radiesthetic reading: 50
Diameter: 17.3 ft
Circumference: 55.1 ft
Height: 108 ft

Additional details: The doorway faces south-southeast and is 14.8 feet above the ground. The windows in the drum in ascending order face west-southwest, northwest, east-southeast, and southwest. The five windows in the story just below the string course under the battlements face north, northeast, east-southeast, southwest, and west-northwest. The conical cap was replaced with castellations from the 1730s. The tower is composed of finished and coursed granite in the lowest 10 feet of the tower, contrasting with less-evenly coursed and smaller local limestone in the rest of the tower. The doorway is of dark red sandstone carved with chevrons, lozenges, and stylized marigolds.

Oughterard Round Tower

Geographic coordinates: 53°16'39" N, 6°33'57" W

Elevation: 452.8 ft

Radiesthetic reading: 38

Diameter: 15 ft

Circumference: 47.2 ft

Height: 31.5 ft

Additional details: The tower is composed of uncoursed spalled limestone, dressed to the curve. The doorway and arched window are of granite. The east-facing doorway has a three-stone arch devoid of decoration at 8.7 feet above ground level. The single window at the second-story level faces south and echoes the design of the doorway.

Taghadoe Round Tower

Geographic coordinates: 53°21'11" N, 7°16'47" W

Elevation: 216.5 ft

Radiesthetic reading: 40

Diameter: 16.3 ft

Circumference: 51.2 ft

Height: 65 ft

Additional details: The tower is made of slatey limestone. The doorway is made of granite, is 11.7 feet high, and faces south-southeast. The three windows face west, south-southeast, and west-northwest.

COUNTY KILKENNY

Aghaviller Round Tower

Geographic coordinates: 52°27'54" N, 7°16'08" W

Elevation: 279 ft

Radiesthetic reading: 50

Diameter: 16.4 ft

Circumference: 50.9 ft
Height: 64.3 ft

Additional details: The tower is composed of slate-colored sandstone coursed and dressed to the curve inside and out. The original northeast-facing doorway is over 13 feet above ground. There is a single square-headed window facing southwest at the second story level.

Fertagh Round Tower

Geographic coordinates: 52°46'42" N, 7°32'41" W
Elevation: 390.4 ft
Radiesthetic reading: 42
Diameter: 15.7 ft
Circumference: 49.2 ft
Height: 102 ft

Additional details: Limestone. The badly defaced doorway is 10.8 feet from the ground and faces northeast. It is fitted with a modern opening, possibly by the OPW in 1879–1880, having allegedly lost the original stones to a farmer who used them as firebricks in his kitchen in the belief that they would protect him from fire. There are five windows in the drum of the tower and four in the top story. Of the lower windows, two are angle-headed and three are linteled, facing south-southeast, north, west, northeast, south-southeast. The four upper windows are angle-headed and face the cardinal compass points.

Kilkenny Round Tower

Geographic coordinates: 52°39'3" N, 7°15'25" W
Elevation: 190.3 ft
Radiesthetic reading: 58
Diameter: 14.8 ft
Circumference: 45.9 ft
Height: 98.4 ft

Additional details: The tower is made of limestone. An excavation in 1846–1847 found that the foundations extend less than three feet down. The arched doorway is 8.9 feet above ground, faces south-southeast, and has no decoration. The windows, all linteled and in ascending order, face northwest, northeast, just to the east of south and just to the south of west. The top story below the parapet has six evenly spaced windows, unique, except for the round tower at Kilmacduagh.

Kilree Round Tower

Geographic coordinates: 53°27'27" N, 6°13'28" W
Elevation: 101.7 ft
Radiesthetic reading: 50
Diameter: 15.9 ft
Circumference: 50.1 ft
Height: 86.5 ft

Additional details: The tower is composed mainly of limestone and has an arched sandstone doorway, which is 5.3 feet above the ground and faces south. Two windows in the drum face the north and east and both are linteled. The top story has four linteled windows facing the compass points.

Tullaherin Round Tower

Geographic coordinates: 52°34'45" N, 7°07'48" W
Elevation: 272.3 ft
Radiesthetic reading: 46
Diameter: 16.1 ft
Circumference: 50.9 ft
Height: 7.38 ft

Additional details: The tower is composed of dressed breccia. The windows are all linteled and face, in ascending order, southeast, west, west-northwest, and north-northeast. The top story, which appears to have had eight windows similar to O'Rourke's tower at Clonmacnoise, has four remaining windows, all rectangular and linteled and facing north-west, north, northeast, and east. The doorway, which has no casing, is supported by a stabilizing pier inserted in 1892 and is 12.1 feet above the ground. The tower is capless and parapeted.

COUNTY LAOIS

Timahoe Round Tower

Geographic coordinates: 52°57'37" N, 7°12'13" W
Elevation: 397 ft
Radiesthetic reading: 43
Diameter: 18 ft
Circumference: 57.4 ft
Height: 96 ft

Additional details: Limestone with some sandstone. The receding door-way is 17 feet above ground level facing east-northeast. The lowest win-dow, facing south, is also unusually finished, being large and framed in stone. The window midway on the rear of the drum is a linteled slit, and the top drum window is squared and linteled, facing west-northwest. The four top-story windows face the cardinal compass points. The tower was renovated by the OPW in the late nineteenth century when the cap was repaired and a modern ground-level doorway was filled on the southwest side of the tower. This is a complete tower without floors or ladders. It has one of the finest four-order Romanesque doorways in Ireland, elaborately carved and decorated with interlace, human heads, chevrons, and capitals. It is unique in round tower architecture.

COUNTY LIMERICK

Ardpatrick Round Tower

Geographic coordinates: 52°20'18" N, 8°31'56" W
Elevation: 715.2 ft

Radiesthetic reading: 25
Diameter: 17.22 ft
Circumference: 54.1 ft

Additional details: This is a stump with dressed and squared stones.

Dysert Oenghusa Round Tower

Geographic coordinates: 52°31'15" N, 8°44'41" W
Elevation: 72.2 ft
Radiesthetic reading: 42
Diameter: 17.3 ft
Circumference: 54.5 ft
Height: ~53.8 ft

Additional details: The tower is made of limestone and the door and windows are made of sandstone. The OPW did some restoration work in 1881–1882. The door is 15.1 feet above ground and faces east. There are three windows that face west, south-southwest, and north-northeast.

Kilmallock Round Tower

Geographic coordinates: 52°24'04" N, 8°34'29" W
Elevation: 282.2 ft
Radiesthetic reading: 35
Diameter: 17.3 ft
Circumference: 54.5 ft
Height: 55 ft

Additional details: The tower was incorporated into the church and is scarcely recognizable as a tower.

COUNTY LOUTH

Dromiskin Round Tower

Geographic coordinates: 53°55'19" N, 6°23'53" W

Elevation: 42.7 ft

Radiesthetic reading: 38

Diameter: 17.1 ft

Circumference: 53.8 ft

Height: 50 ft

Additional details: The tower is composed of slatestone. The OPW did some restoration work in 1879–1880 to the already squat tower. The door is 12.1 feet high and faces east-southeast. There is one window that faces west-northwest and four recent windows at the top.

Monasterboice Round Tower

Geographic coordinates: 53°55'19" N, 6°23'53" W

Elevation: 42.7 ft

Radiesthetic reading: 38

Diameter: 16.3 ft

Circumference: 15.3 ft

Height: 91.9 ft

Additional details: The tower is made of slatestone. The east-facing, arched doorway is made of sandstone and is just 6 feet above ground level and framed by a wide double-banded molding. Aside from the beautifully crafted quoins, there is no decoration. Like the doorway, the angle-headed window directly above it is composed of sandstone. The three other windows in the drum are linteled and small and face, in ascending order, west, south, and north. Apparently it is still equipped with floors and ladders, but while there is a concrete and metal stairway to the door, the iron grill fitted into the doorway is padlocked.

COUNTY MAYO

Aghagower Round Tower

Geographic coordinates: 53°55'19" N, 6°23'53" W

Elevation: 42.7 ft

Radiesthetic reading: 38
Diameter: 16.4 ft
Circumference: 51.7 ft
Height: 51.7 ft

Additional details: The tower, composed of limestone, was rebuilt in 1969. The doorway is 7.2 feet above ground and faces east. There are three windows that face south-southwest, west-southwest, and south. There is no roof. The modern ground floor doorway allows entry into the tower.

Balla Round Tower

Geographic coordinates: 53°48'18" N, 9°7'53" W
Elevation: 124.7 ft
Radiesthetic reading: 27
Diameter: 17.3 ft
Circumference: 54.3 ft
Height: 32.8 ft

Additional details: This abbreviated tower is composed of sandstone and has two doorways. The upper and possibly original doorway is placed higher than almost any other round tower at 26 feet above the present ground level. There is one tiny arched window on the south side of the tower, perfectly carved using four stones, about halfway between what would have been the first and second floors of the tower.

Killala Round Tower

Geographic coordinates: 54°12'56" N, 9°13'14" W
Elevation: 32.8 ft
Radiesthetic reading: 55
Diameter: 16.5 ft
Circumference: 51.8 ft
Height: 84 ft

Additional details: The tower is made of limestone and has some very large stones. The round doorway is 12.5 feet high and faces south-southeast. There are three square windows in the drum that face, in ascending order, east-northeast, south-southeast, and west. The traditional four windows in the top story are square-headed internally and angle-headed externally and do not face the cardinal compass points, but are slightly skewed, facing north-northeast, east-southeast, south-southwest, and west-northwest. Repairs were conducted in 1841, and then by the OPW in 1882–1883.

Meelick Round Tower

Geographic coordinates: 53°55'17" N, 9°01'2" W
Elevation: 111.5 ft
Radiesthetic reading: 37
Diameter: 17.9 ft
Circumference: 56.1 ft
Height: 69.9 ft

Additional details: The tower is composed of quartzy sandstone with some limestone. The five-stone arched doorway is 11.2 feet above the ground and faces south-southeast. Of the six windows, the lowest two are angle-headed facing east-northeast and south, and the four remaining linteled windows face north-northwest, west-southwest, east-southeast, and north. There is no cap or top story. OPW did some repointing work in 1880–1881, but the details are not recorded.

Turlough Round Tower

Geographic coordinates: 53°53'18" N, 9°12'30" W
Elevation: 111.5 ft
Radiesthetic reading: 50
Diameter: 18 ft
Circumference: 57.4 ft
Height: 75 ft

Additional details: The tower is composed of quartzy sandstone with some limestone. The arched doorway (currently blocked with mortared stone) is 13 feet above the ground to the southeast. Angle-headed top-story windows face just to the left of the cardinal compass points. The other small linteled windows in the drum from bottom to top are oriented to the south, then slightly skewed to the west, north, and east. The tower was repaired in 1880 by the OPW.

COUNTY MEATH

Donaghmore Round Tower

Geographic coordinates: 53°40'13" N, 6°39'44" W
Elevation: 167.3 ft
Radiesthetic reading: 68
Diameter: 16.3 ft
Circumference: 51.3 ft
Height: 81 ft

Additional details: The tower is made of limestone and has a rounded sandstone east-facing doorway, which is about 11.2 feet above ground. Above the door is a badly weathered image, possibly a human torso with arms outstretched, contested fiercely by George Petrie as being a crucifixion and part of the original structure, whereas John Windele, a nineteenth century scholar often quoted by Petrie, claimed it was a later addition. The stone in question is laid with positive magnetic polarity upwards, reinforcing Windele's opinion. There are four windows which are, in ascending order, a south-facing linteled window, an angle-headed east window, an arched west window, and another linteled window just below what would be considered the top story. The tower was reportedly repaired by an owner in 1841 and some restoration work was also done by the OPW at a later date.

Kells Round Tower

Geographic coordinates: 53°43'8" N, 6°52'48" W
Elevation: 275.6 ft
Radiesthetic reading: 55
Diameter: 16.2 ft
Circumference: 50.9 ft
Height: 85.3 ft

Additional details: The tower is composed of limestone. The round-headed doorway is 10.8 feet above the offset, but due to an earth-filled base, 6.1 feet on the other side, and faces north. There are four small linteled windows in the drum that face, from bottom to top, south-southwest, east-southeast, north, and west-northwest. The top story windows of the tower are unusual in that there are five of them rather than the usual four, with Kildare being the exception. They are evenly spaced and face east-northeast, southeast, south-southwest, west, and north-northwest.

COUNTY MONAGHAN

Clones Round Tower

Geographic coordinates: 54°10'39" N, 7°13'58" W
Elevation: 190.2 ft
Radiesthetic reading: 42
Diameter: 16.3 ft
Circumference: 50.5 ft
Height: 50.5 ft

Additional details: The tower is made of sandstone. The east-facing doorway is 5.4 feet above the present level and 7 feet above the offset. Small linteled windows face, in ascending order, south, north, and east with the traditional but probably late four top-story windows at the cardinal compass points.

COUNTY OFFALY

Clonmacnoise Round Tower

Geographic coordinates: 53°19'35" N, 7°59'11" W
Elevation: 128 ft
Radiesthetic reading: 68
Diameter: 18.4 ft
Circumference: 57.9. ft
Height: 63.3 ft

Additional details: The tower is made of limestone. The doorway faces southeast at a height of 11.5 feet above the offset at the base. It is a slightly triangular-shaped arch, being narrower at the top, wider at the bottom. The arch is beautifully cut of nine matching stones resting on transitional stones that project slightly into the doorway and above the five jambstones on either side. All stones carry through the entire depth of the wall. There are ten windows. One linteled window to the northwest is one level above the doorway. Another linteled window is to the north facing the River Shannon. The other eight windows, all linteled, face the cardinal compass points and opposing points between.

COUNTY SLIGO

Drumcliffe Round Tower

Geographic coordinates: 54°12'56" N, 9°13'14" W
Elevation: 32.8 ft
Radiesthetic reading: 55
Diameter: 16.3 ft
Circumference: 51.3 ft
Height: 28.9 ft

Additional details: The east-southeast-facing doorway is 5.7 feet above ground level. One very small linteled window survives facing south-southwest at about the second floor level.

COUNTY TIPPERARY

Cashel Round Tower

Geographic coordinates: 52°31'13" N, 7°53'25" W
Elevation: 100.3 ft
Radiesthetic reading: 57
Diameter: 17.4 ft
Circumference: 55 ft
Height: 91.7 ft

Additional details: Excavations done in 1841 found this sandstone tower built directly on the solid rock outcropping. The original doorway faces southeast, 10.8 feet above ground. The arch is composed of seven stones with five stones in the west jamb and six in the east. There are three square-headed linteled windows in the body of the tower and four angle-headed windows in the top story, off the cardinal compass points in the northeast, northwest, southeast, and southwest. Two of these angle-headed windows are formed with a single stone, unlike the more common configuration of two stones set to either side forming a 90-degree angle.

Roscrea Round Tower

Geographic coordinates: 52°57'21" N, 7°47'45" W
Elevation: 315 ft
Radiesthetic reading: 68
Diameter: 15.1 ft
Circumference: 50 ft
Height: 65.6 ft

Additional details: The tower is made of limestone and has no cap. The arched, south-southeast-facing doorway is just about three meters above the present ground level. The large east-facing window in what would have been the second story is angle-headed. It is notable for the single-masted ship carved in relief on the north jamb toward the inner edge. There are two other windows in the drum of the tower. Both are much smaller and linteled, with the lower of these facing west and the upper facing north.

COUNTY WATERFORD

Ardmore Round Tower

Geographic coordinates: 51°56'54" N, 7°43'34" W
Elevation: 121.4 ft
Radiesthetic reading: 72
Diameter: 16.4
Circumference: 51.8 ft
Height: 85.3

Additional details: The tower is composed of sandstone. The door faces east-northeast and is 12.4 feet above ground. There are seven windows that face north, east-northeast, and south-southwest, four top-floor windows face the cardinal points. The tower also has the unique feature of three string courses, which are rounded and project in an unbroken circle approximately 20 feet above ground level for the first course, 36 feet above ground level for the second course, and about 57.4 feet above ground level for the third course. Five of the sixteen corbels are uniquely carved.

COUNTY WICKLOW

Glendalough Round Tower

Geographic coordinates: 53°00'38" N, 6°19'40" W
Elevation: 452.8 ft
Radiesthetic reading: 72
Diameter: 16 ft
Circumference: 50.2 ft
Height: 100 ft

Additional details: The tower is composed of mica schist and some granite and received a new cap in 1876. The door faces south-southeast and is 10.5 feet above the offset. There are four windows that face southwest, northwest, north-northeast, and east-southeast. There are four larger windows at the top that face north-northeast, east-southeast, south-southwest, and west-northwest.

SCOTLAND

Abernethy Round Tower

Geographic coordinates: 56°19'58" N, 3°18'42" W
Elevation: 121.3 ft
Radiesthetic reading: 37
Diameter: 15.2 ft
Circumference: 48 ft
Height: 72.2 ft

Additional details: This tower is made of sandstone and has a north-facing door 8.2 ft high, three windows facing south, west, and east, and four larger windows at the top on cardinal points.

Brechin Round Tower

Geographic coordinates: 56°43'50" N, 2°39'41" W
Elevation: 173.9 ft
Radiesthetic reading: 37
Diameter: 14.9 ft
Circumference: 46.9 ft
Height: 85.2 ft

Additional details: The tower is made of sandstone. The arched door faces west, is 7.2 feet above ground, and is elaborately carved with a tau cross. There are two windows in the barrel facing east and south and four at the top facing cardinal points, with later additions above.

■ ■ ■

There seem to have been twenty-three towers that have been demolished or destroyed, or are at least no longer there, but reference has
been made to them at some stage in the past. Details about them can
be found in Barrow's *Round Towers of Ireland* or Lalor's *Ireland's
Round Towers*.

9

The Purpose behind
the Round Towers

I t is a commonly held belief that stasis in any environment—whether it is in a human organism or a setting in Nature—causes issues that if not removed result in problems. The Chinese, for instance, in their traditional medicine, go to great efforts to restore the movement of nerve, blood, and other essential fluids to avoid blockages in passages that must be kept open if good health is to be enjoyed.

By observing Nature over time, humans have accomplished the most remarkable feats not only to improve living conditions for ourselves and our domestic animals but also to enable a healthier environmental state. Such feats include removing swampland to reduce the mosquito population, reclaiming the desert by planting trees to retain water in the soil, and so on.

In a country like Ireland where water is to be found everywhere, both above and below ground, any method used to enhance its management would be welcome, especially when it causes damage. Of course, the next questions are "What damage?" and "How do we even know that this damage is caused by the flow or the crossing of underground water?" There is no ready answer, but given the feedback we have accumulated over time from hard experience and the knowledge base—for the believers—as seen in chapter 6, it seems quite clear that we can choose to believe that option, or not.

Modern skepticism is a great boost for science and probably the force behind many discoveries, but at some stage one has to take a position and believe. I personally found that belief when I lost my sight due to the influence of an underground stream, and in the subsequent search for a solution, that influence was removed and my eyesight was restored.

There is no way I can convince everyone, and I have absolutely no desire to do so, but my respect for the wisdom of our ancestors and our lack of understanding in that regard encourage me to at least give them the benefit of the doubt. And what better way can I repay such a debt, in this modest manner, than by acknowledging the contribution of those anonymous benefactors.

If there is no branch of science that deals with such phenomena and the multitude of accompanying side issues of telluric forces, there is little surprise that any attempt to discuss them is met with contempt and the accusation of pseudoscience. If researching ancient lore, current problems, and branches of neglected knowledge is pseudoscience, I *am* a pseudoscientist and proud to be so. That might just be one way of bringing some relief to the suffering of Nature.

Unfortunately, access to the majority of "scientific" studies into geopathic stress has to be paid for. Elsevier and JSTOR seem to have a fair volume of studies that could well help us in any effort to learn, but experimentation and research in feng shui and vastu produce practical solutions rather than theoretical propositions, which are more in keeping with what I believe to have been the spirit and societal concerns of our ancestors.

What I suggest is that there is a strong chance that when all the towers were in their original state, complete with their caps and operating at full magnetic and energetic capacity, they were fulfilling a task the effects of which are still felt to this day in the immensely rich agricultural land that the island offers. What is more, the feel-good ambience found in Ireland is perhaps due, to a certain degree, to these intriguing structures.

It is pretentious no doubt even to put forward a theory as to what the towers are for, but as you have read above, there have not been any convincing arguments put forward in the past, and given what we can imagine must have amounted to a complex and highly laborious task, we can reasonably assume that not only was there a motive for building the towers but also that it was worth the effort.

There can be little doubt that some of the most convincing writing on the purpose of the round towers comes from the aforementioned Philip Callahan—polymath, entomologist, biophysicist, infrared- and low-energy expert, traveler, and immensely curious and avid observer of Nature. In several of his books—*Ancient Mysteries, Modern Visions, Exploring the Spectrum,* and *Paramagnetism,* he explains that the most practical aspects of the round towers are their operation as wave receivers and transmitters and their ability to boost agriculture.

I would add a detail that he does not address: the *how.* How did our ancestors achieve such a boost? Is it possible to achieve the necessary drawing capacity, as in the magnetic quality, to pull up that cocktail of frequencies developed by the two or more currents, and other possible factors from below ground?

The answer seems to lie in the fact that every stone in each and every round tower has been deliberately placed in a specific position by the mason. As discussed in previous chapters, the stones have been systematically laid with the negative magnetic polarity facing upward, always on the outside walls of those towers that have not been tampered with and most probably on the inside, but it is generally impossible to access the inside of a tower to check that.

If Callahan had suggested that the towers were operating as an energy-resonating system derived from the water below ground rather than from the ether above, then we would really have a working hypothesis. He didn't, but I do.

It seems timely now to try to pull all the aspects of what I believe into a coherent story to explain what the round towers were used for and how they worked.

THE PURPOSE AND FUNCTION
OF THE ROUND TOWERS

The flow of water, like any movement or current, generates a magnetic field. Over time, that field can expand if able, thus affecting the surroundings with the frequencies produced. Due to water capillaries or fissures in the bedrock or soil—or whatever the geological medium found between those currents and the surface is—those frequencies, if unable to find an exit below ground, will eventually rise to the surface and exit the earth.

In all probability, this sets up an upward pathway to be followed by other frequencies as they develop, resulting in a substantial pattern in which frequencies are continuously being released into the environment, where they may or may not be of benefit for the surrounding flora and fauna, as they slowly rise into the atmosphere. Because there is some form of disharmony or a lack of resonance, enabling assimilation with the frequencies coming from below, that force is at its most intense at the exit point; hence the damage that can be worked unless measures are taken to resolve it. It would be reasonable to suppose that there is a very delicate balance in Nature that has taken millions of years to establish, and any upset in that balance can cause problems for some and perhaps provide benefits for others. However, we are at a total loss of appreciation either way.

As explained above in chapter 6, one of the most harmful combinations of energies is created when two or more currents of water cross below ground because they create a cocktail of frequencies that we normally don't deal with and are not equipped to do so. When they reach the surface and work their particular effects, they upset the normal equilibrium, and disease generally follows.

However, when that frequency cocktail is elevated beyond our immediate vital space, it is as if we are left in a vacuum, where we are relatively free from the interfering frequencies, and there is room for the individual energetic composition to express itself not only with less restriction but in harmony with the surroundings. That neutrality is very much what one is aiming for when removing the frequencies from the area of the exit point. Again, in all probability conditions in our environment, such as temperature, daylight, pressure, the state of ionization, and other factors all play their part in the overall result. Consequently, if one can operate on one or more of those factors, there is a chance that a comfort zone can be achieved.

Because of the layering effect of the oriented stones, the barrels of the towers, like the stone pillars of the cathedrals, are able to concentrate and focus that energy from below, carrying it upward and releasing it through the doorway and the windows on the way. Consider how much more efficient a double hollow barrel would be compared to a solid pillar, especially if you arranged a polarized array of stonework like that in the round towers to draw that magnetic force field up so that as it bursts through the door and windows, it works its modified and beneficial effect on the environment in the whole area, giving it a quite unique quality.

As long as a tower is located over the crossing of underground water, it does not matter what type of stone is used, but all of the stones must be laid with the negative magnetic polarity facing up, so that they create the necessary magnetic thrust to draw those frequencies up and away. In all probability there is a connection among underground waterways, thus enabling a form of pressure release that allows the water to stay closer to the surface, giving the Emerald Isle its unique properties.

The arrangement of the doorways and windows in the towers would appear to create a spiraling movement, the force of which is released through the openings and in the process somehow seems to remove the harmful effect of the combined frequencies from below as the "pressure" is released. Is there a lumination (interaction of external environmental energy and the individual's energy field, as per Dr. Wilhelm Reich) effect operating, transforming that magnetic field into a negative ion complex? This is where the unknown components lie and the magic is worked. Whatever that effect is, it is going to have an impact on the surrounding area and especially on the more sensitive components, such as the mycorrhiza.

One can easily imagine the effect of concentrating a source of energy such as the magnetic field of force coming from the crossing of underground water by allowing that field to rise through the barrel of a tower, which narrows and narrows again four or five times on its way up to the top, thereby further focusing the force. By releasing the accumulated energy through the doorway and windows, the purpose of the energy flow is fulfilled and maintained at a constant rate. Such an arrangement would militate for a hole in the roof of the tower, as a cap with a hole in it would provide for a more consistent release, ensuring a constant and more regular flow of that force than would a sealed cap. Or perhaps the windows in the top story accomplished that. It would not be difficult to perform some simple scale experiments using an energy-generating source, as I have done with a geometric spiral as the source and a series of five copper cones of 3-inch diameter spaced regularly.

The fact remains that whatever we call "energy" is in and around us all, and it is modified by every event, however small or insignificant, but to what extent or depth shall remain an unknown until the answer either hits us in the face, or we knowingly devise some form of method or other to answer the question. If we unconsciously control our existence by breathing, digestion, and circulation, how much better a job could we do of it if we were to assume control of the few factors that we can regulate? Perhaps that's just what the ancients were doing! You know the benefits you can achieve by breathing or relaxing muscular tension because of the positive impact it has on emotions and thoughts, so why not apply those same principles by means of a round tower?

In the 1930s, Wilhelm Reich (1897–1957) was able to objectively measure the movements of a bioelectric charge that he named "orgone" by using a very sensitive millivolt meter with sensors attached to the body to record the subtle bioelectric charge. He found that the energy flowed from the inside body core to the outside surface (toward the world) when a person felt pleasure or expansion, and it flowed from the surface to the interior (away from the world) during states of anxiety, fear, and contraction. There seems to be room for somewhat of an analogy between the energies coming from the earth when released into the atmosphere (see, for instance, lumination above).

Reich also noted that the conditions of expansion and contraction affected a person not only emotionally but down to the autonomic nervous system, to the cellular and even chemical levels. States of expansion produce parasympathetic conditions associated with the dilation of the blood vessels and increased circulation, including pain relief, better digestion and peristalsis, lower blood pressure, and the stimulation of potassium and lecithin production and also creates sexual excitement and a sense of well-being.

States of contraction, however, produce sympathetic effects, including constricted blood vessels, less blood flow, and often pain. In addition, the contracted condition increases blood pressure and heart rate, adrenaline flow, and cholesterol. It inhibits digestion and blood supply and is associated with the emotions of anxiety and stress. The ability of the body to expand and contract and not become "stuck" in one mode created what Reich called the "pulsation of life," which distinguished the living from the nonliving. This pulsation of expansion and contraction also followed a specific rhythm.

If we are ever to resolve the purpose of the round tower question once and for all, I think we would need to manufacture some simple instrumentation that is capable of measuring the magnetic force field at (1) the exit point of an underground water crossing, (2) in the center of a tower, (3) at the doorway, (4) at the windows, and then at varying distances from the tower. While I have taken those measurements using a pendulum and a chart of my own making, these methods will not in all likelihood satisfy the scientific-minded.

◇ *Method for Taking Pendulum Readings*

As a guideline for anyone wanting to repeat these readings with a pendulum, here is the procedure I employed.

1. Having located the crossing of underground streams, physically or remotely on a map, ask what is the score? For an honest, as the French dowsers say when referring to the simple as opposed to combined, effect of a telluric event, the most powerful reading I have found is –50.

2. It is rarely easy to access a tower to take this measurement. I only did it once at Aghagower in County Mayo, and was rewarded to discover a score of +70 inside, compared with +50 outside of the precincts of the tower. That was also the occasion to check the alignment of the stones of the inside wall.

3. Some towers allow for ready measurement at the doorway, as they are sufficiently low down and accessible, others require projection of your dowsing skill to the height and location of the door. The measurement of the doorway of Kilmacduagh I found to be +85, and at Kinneigh, +60.

4. Measurements at the windows show an average score of +60.

5. It is then of interest to find out how far the tower effect carries. This requires a fair amount of legwork, which is not always feasible due to dense graveyards, church buildings, enclosures of varying sorts, roads, homes, gardens, and so on. Where possible, I would walk on a rough compass bearing for 500 yards or so, taking measurements every 50 paces or so. As to be expected, the strength of the reading diminishes with distance. It is not uncommon, however, to find a reading of +20 with the tower in sight even at a distance of a mile or more—Kinneigh, Aghagower, Kilmacduagh, Killala, and Ardmore, for example.

■ ■ ■

Having laid out the case for your consideration, it is time to go into greater detail about the various principles that the ancients seem to have mastered and applied.

THE SECRET SCIENCE OF THE TOWERS

We will now take a closer look at the evidence we still have at our disposal, and hopefully in the light of what has been explained above, make it easier to consider—perhaps even understand—the methodology employed in the round towers. In a manner of speaking, we have to work backward and reverse engineer the how and why because we have no user manual for this phenomenon we seem to be dealing with.

If you have ever visited any of the cathedrals built in western Europe during the twelfth or thirteenth centuries, you will have perhaps experienced a unique sensation of calm that is not found in any other ecclesiastic structure. A sensation on walking into the building that a weight has been removed from your shoulders. This, I believe, is due to the use of the unique architectural principle of measurement, or the geographic principle as I call it here, and as explained above.

Vibration is a phenomenon caused by the motion of periodic or random frequencies generated by an energy source, for instance insect antennae, emotions, the heart, one's voice, electric (infrared) electromagnetic impulses, sounds, and so on. A less obvious but equally potent source of vibration is architecture, with its volumes, proportions, and consequent projections. This is where Callahan could well have been close to the truth, but rather than outward-facing antennae, the towers are inward-facing—to what is below the ground.

Whatever form the vibration takes, the movement naturally enough interacts with the surrounding environment and especially the human complex as it is composed of zillions of cells all vibrating at as many frequencies, or so we are told. One can assume that the resulting chaos is uncomfortable and any relief provided by whatever means might be welcome, but I digress.

What if the heart's beat did more than keep the blood flowing? What if it was the key to human, indeed all life-forms with a heart? The heartbeat might be an electric impulse originating in the combination of radiation from the cosmos, the earth, and the environment.

Or perhaps we are ambulant thermoelectric stations with the heart, producing a constant vibration; a resonant cavity pulsating thanks to its expansion and contraction and perhaps the conduit for a host of information coming from the outside via contact with the skin and the senses, both receiving and transmitting. It does not seem an unreasonable proposition as it clearly does more than keep our life blood circulating. We know that an antenna does not have to be made of metal; both insulative (dielectric) and conductive materials can fill that role, so why not a heart inside a resonating bundle of energy?

No matter the explanation, our actual experience provides the best, and only, guide. When we awake from sleep, we move out of that state of blissful restoration—so long as we do not sleep over geopathic stress—found in deep sleep or in conscious immersion into what is known as the parasympathetic nervous mode. The waking state is dependent on the sympathetic nervous mode—the flight or fight response—which brings a number of reactions with it, such as activating the brain, the muscles, the pancreas, the thyroid, and the adrenals while at the same time stimulating certain hormones and substances, including insulin, cortisol, dopamine, serotonin, the thyroid hormones, and a host of others.

While we are very good at differentiating and defining these substances, we do not know much about the states, reactions, and effects generated by their interaction with other substances and with our environment. This is a great loss because we send out vibrations reflecting our internal condition, which could potentially have an impact on everything around us, and in turn there is a knock-on effect on the environment, subsequently feeding back to ourselves. We could learn much from these interactions, if only we were truly in tune with ourselves—and our environment—and this is what I think the ancients were driving at with their stone-tower science.

While on the subject of vibration, there seems to be a commonly experienced isolation felt in certain places throughout the world. This is an inevitable consequence of the buildup of all sorts of external vibrations in the form of electromagnetic radiation, especially noticeable in shopping malls and motor vehicles. Winfried Otto Schumann was the first to study the theoretical aspects of resonance from the earth's ionosphere, although an Irishman, George Francis FitzGerald, had put forward the idea in 1893 that the top layers of the atmosphere are good conductors of electromagnetic waves. It is a recognized fact that anyone traveling in space suffers from jet lag and disorientation, the reason being that in the ionosphere and beyond there are no longer the same waves we enjoy here below on the earth. When humans are deprived of these frequencies because they are no longer connected with the earth, their metabolism is upset. In the same way, when you visit a shopping mall you are surrounded by an intense curtain of electromagnetic radiation coming from every which way and from multiple sources, which deprives the atmosphere of beneficial negative ions and supercharges it with harmful positive ions, which can cause fatigue, brain-fog, disorientation, and irritation.

We live within an electric field (well, that's how we choose to term it) that varies according to climate, location, and individual height, with a negative charge at ground level potentially rising by 100 volts every 3 feet (up to higher altitudes, when it drops off again). So, at 6 feet tall you have at your nose's disposal 200 volts; this is the earth's naturally occurring field.

In a shopping mall you are deprived of beneficial negative ions but immersed in an atmosphere supercharged with harmful positive ions. Similarly, when you are in a car—which is a metal box, or Faraday cage—you are disconnected from the grounding capacity of the earth, more or less, because of the rubber tires, and as a result are subject to road rage, greater fatigue, and other effects because you are no longer in contact with the potential provided by the earth's frequencies.

Now, what if one combines this understanding with a building technique that can harmonize vibration as I believe the technology of the stone towers was designed to do? The rediscovery of such a technique has been written and speculated about, albeit obliquely, by Louis Charpentier in his book *Les Mystères de la Cathédrale de Chartres* (*The Mysteries of the Chartres Cathedral*), and what a fascinating intrigue it all makes. Basically, the story runs as follows:

Saint Bernard de Clairvaux (1090–1153), a Cistercian (the monastic order renowned for their affinity to water), apparently learned of a secret concealed in the Temple of Solomon in Jerusalem (the British were still digging in the temple during their occupation between the world wars). Quite what the secret was, history does not tell, but Charpentier tends to believe it was related to the Ark of the Covenant. Whatever it was, Bernard rapidly set up (in association with Hugues de Payens) a monastic-soldier order of very influential nobles from the Champagne region to bring the secret(s) back to France. It must be remembered that all the knight members of this order were of "noble" birth (i.e., families of influence due to their financial and political clout). Together they were known as the Knights Templar. Fascinating though the plot may be, it is not our concern here, and the secret sank even deeper with the disbanding of the Templars in 1309.

There was, however, the most remarkable consequence of the Templars being established and the impact of the "secret" they brought back to France (or what we know as France today). For a period of almost two hundred years, from about 1130 to 1310, a Europe-wide cathedral-building campaign was maintained, with construction in France, Germany, Portugal, Spain, and the UK. In France alone, one hundred and fifty churches were commenced between 1150 and 1250. It was the most rapid and widespread construction program of cathedrals ever seen in the history of man, and what is more, it was accomplished using a totally unique style of architecture, the so-called Gothic style, with innovative techniques never employed before or since. The extent, intensity, and precision of building has never been repeated.

It is no easy task to grasp what a huge undertaking the building of a cathedral was. Even when you are in the center of the labyrinth inside Notre-Dame de Chartres, looking up at the roof, the immensity and the overwhelming sensation it elicits tend to make you feel very small and prompt a rather special mindset and/or imagination.

A period of twenty-five to forty-odd years is needed to build such an intricate structure. Even with a company of specialists and an army of laborers, that is not a long time, and the task is a logistics nightmare with people who need to be paid, housed, fed, and organized with clockwork precision, not only because their services are costly but also because they are required to be on other cathedral worksites—fifteen in twelfth-century Normandy alone.

One must reasonably assume that the Templars had access to plenty of cash, thanks to their international network perhaps. The Roman church was rich even then, but not that rich, and very probably could not have financed all of the construction sites. Once again, rumors abound, such as silver from South America traded for wool, or a new economic model, not to mention the political considerations, for there seems to be a considerable probability that there was a major conflict between the church and the Templars, which the Templars ultimately lost. One of the consequences of that struggle concerns us directly here, inasmuch as the loss of this architectural principle seems to have disappeared with the Templars.

It is now time to return to our original quest, for this incident is but a stone in the path, albeit a major stepping stone bathed in mystery, setting the tone for the death struggle that we do not hear about in polite society when mention is made of the church. Once more, it is the victors, the Roman church again, who have written a history which conceals so much of this troubled medieval period. The Roman church authority was put to serious trial between the eleventh and fourteenth centuries, firstly by the Cathars, a grass-roots rather than the noble-led Templar movement. The Cathars were firmly crushed in the Albigensian crusade with the loss of a million lives or so in the southwest of France, whereas the Templars gradually disappeared from Scotland and Portugal whence they had escaped for refuge.

Whatever the truth, the cathedrals were erected across Europe, including the UK, during this period—*all* using the same Geographic Principle discussed in chapter 7. Architecturally speaking, while the Gothic and Roman styles appeared to be concomitant and continued over the next few centuries, never again was this Geographic Principle, found uniquely in the structures built during the Templar period, to be used in any form of architecture after the disappearance of the Templars. The order was disbanded in 1309, and the last master, Jacques de Molay, was burned at the stake in 1307 on the orders of Philippe le Bel, the king of France. Construction of the great cathedrals ceased in Europe around that time.

Jacques de Molay, prior to his horrendous ordeal, had spread the word that the soldier-monks would be received with open arms in Scotland. (Portugal of course became home to many of them, and Freddy Silva in his excellent book, *First Templar Nation*, explains that all very well.) The fortunate few who did escape did not apparently include any of the brethren who were privy to the building secrets. The reason for saying this is that the Rosslyn Chapel, perhaps one of the finest works of art of a newfound masonic society, does not display either the Geographic Principle or the Polarity Principle. And although the building is located over the crossing of underground water, such a location is common to all ancient sites that were usurped by the new arrivals, i.e., the church.

This brings us full circle back to the idea proposed at the beginning of this book, namely the contrived disappearance of the Druids and their way of interacting with Nature, based (like other traditions such as the Daoists) on careful observation. This mindset was the very antithesis of immediate and total control expressed by the materialistic Romans. The gentle ways developed by contemplation of one's observations are the first victims of totalitarian domination, and so it would seem that those occulted principles learned and, fortunately applied in the towers rather than being completely lost to our view, are once more raised from obscurity. It is now for us to develop a true science of magnetism and continue the extraordinary efforts of our ancestors for the benefit of Nature and all her denizens.

Conclusion

(If such a thing is possible)

The search for truth is obviously a very subjective matter, and in the final analysis, a very egocentric one because we are all determined to be right, and to a certain extent we are. It is the extent, or the scope, that is judged by others and over time determines the validity.

Obviously, the truth with regard to a subject can only be found when all evidence is collected and assessed in the context of that subject. If any aspect is uncertain, out of place, or inconsistent with the evidence and cannot fit into the original context, it should be considered as such and afforded the merited judgment. If all aspects of a subject are considered and found to be in harmony, it can be considered as probably right and as close to the truth as we can determine with our parameters limited and biased by time and context. Hopefully, thanks to a holistic approach of thorough investigation of what is accessible, we can determine what is right or otherwise, leaving to one side human opinion, and much more, our tendency to deride when we do not agree with what others suggest.

A second point, and perhaps the most important in this book, concerns how we assess the evidence, having achieved and registered awareness and understanding of something. In this instance, an awareness of something invisible, subtle, and for which no instrumentation, even if it existed, can prove how it impacts the human ability to know, or gnosis.

As a consequence, when confronted with a situation where it is not possible to know the causes, rather than falling back on opinion (generally that of others), I can only ask that you look impartially at the empirical *effects* so as to link any possible causes in view of what is evidentially available. Furthermore, I would ask you to consider the possibility that kinesiological methods (radiesthesia, in my case) are valid in determining not only the existence of these hidden forces, but also their effect, thanks to an ability to employ our intellectual capacity, experience, and intuition. I am not asking you to cast your cerebral logic and intelligence to the winds but simply to consider the possibility of the existence of such methods in helping us discover the nature of these subtle forces in light of their effective impact on our subject here and maybe on all that we know as life.

Chiang Mai Stone Readings

These readings are taken from experiments conducted in Mae Rim, Thailand on November 20 and 21 on a piece of clear land in the shade of tamarind trees. The average temperature was approximately 68°F and there was an atmospheric pressure of 1,900 or 28.8. The reading of the location was 6,500 on the Bovis scale and the color vibration was green. The rocks used were taken from Wat Umong, probably from a local river, and the pendulum said they are in the region of four hundred and fifty thousand years old. Note that all readings are in Bovis units.

Formation: Square with four right angles
Placement: Four stones positioned at an equal distance of 5.3 ft from each other with positive side facing up, A-B facing north, B-C facing east, C-D facing south, D-A facing west. A: 15,000; B: 15,500; C: 15,500; D: 15,000
Readings:
- Center: 26,000
- Color: Green

Formation: Square with four right angles
Placement: Four stones positioned at an equal distance of 5.3 ft from

each other with positive side facing up, and A-B facing east. A: 15,000; B: 15,500; C: 15,500; D: 15,000

Readings:
- Center: 26,000
- Color: Green
- 131 ft from center: 26,000 in all directions
- 164 ft from center: 25,000 in all directions
- 230 ft east: 22,000
- 262 ft east: 20,000
- 295 ft east: 16,000
- 394 ft east: 8,000

Formation: Square with four right angles
Placement: Four stones positioned at an equal distance of 5.3 ft from each other with positive side facing up, A-B not facing north, and A inverted. A: 15,000; B: 15,500; C: 15,500; D: 15,000

Readings:
- Center: 20,000
- Color: Blue
- 16 ft south: 16,000

Formation: Square with four right angles
Placement: Four stones positioned at an equal distance of 3.2 ft from each other with positive side facing up, and A-B facing north. A: 15,000; B: 15,500; C: 15,500; D: 15,000

Readings:
- Center: 15,000
- Color: Blue
- 164 ft south: 11,500

Formation: Square with four right angles
Placement: Four stones positioned at an equal distance of 3.2 ft from each other with positive side facing up, A-B facing north, and A inverted. A: 15,000; B: 15,500; C: 15,500; D: 15,000

Readings:

- Center: 9,000
- Color: Blue

Formation: Equilateral triangular

Placement: Three stones positioned at 3.3 ft distance from each other with positive side facing up. A: 15,000; B: 15,500; and C: 15,500

Readings:

- Center: 27,500
- Color: Green
- 164 ft south: 27,000
- 394 ft east: 15,000

Formation: Isosceles triangle

Placement: Three stones with A being positioned perpendicularly at 2 ft from B-C. B and C at 3.3 ft distance from each other, positive side up. A: 15,000; B: 15,500; and C: 15,500

Readings:

- Center: 31,000
- Color: Green

Formation: Circle with a radius of 5.3 ft

Placement: Thirteen stones of an average 10,000 with positive side facing up

Readings:

- Center: 34,000
- Color: Green

Formation: Circle with a radius of 5.3 ft

Placement: Thirteen stones of an average 10,000 with positive side facing up and one stone of 15,000 in the center with positive side facing up

Readings:

- Center: 45,000
- Color: White

- From center: 45,000 in all directions.
- 66 ft east: 18,000

Formation: Circle with a radius of 5.3 ft
Placement: Thirteen stones of an average 10,000 with positive side facing up and one stone of 15,000 in the center with negative side facing up
Readings:
- Center: 25,000
- Color: White

Formation: Circle with a radius of 5.3 ft
Placement: Nine stones of an average 10,000 with positive side facing up, one stone of 15,000 in the center with positive side facing up, and four stones positioned at an equal distance of 3.2 ft from each other with positive side facing up.
Readings:
- Center: 51,000
- Color: White
- 164 ft south: 51,000
- 394 ft south: 27,000

Formation: Circle with a radius of 5.3 ft
Placement: Thirteen stones of an average 10,000 with positive side facing up
Readings:
- Center: 31,000
- Color: Green

Formation: Circle with a radius of 5.3 ft
Placement: Eighteen stones of an average 10,000 with positive side facing up and thee stones of 15,000 in the center with positive side facing up
Readings:
- Center: 90,000
- Color: White

- 164 ft south and east: 90,000
- 394 ft south and east: 45,000
- 492 ft south and east: 30,000

Formation: Circle with a radius of 5.3 ft

Placement: Eighteen stones of an average 10,000 with positive side facing up, two stones of 15,000 in the center with positive side facing up, and one stone of 15,000 with negative side facing up

Readings:

- Center: 45,000
- Color: White

A "hot spot" 144 ft from the experimenting area, normally reading 2,000, rose to 22,000 under the square four-stone (5.3 ft) influence. When placing the stones with the negative polarity facing up, the increase in the score of the average reading was 50 percent.

Acknowledgments

Where does one start? So much patience, kindness, and sympathy from family, friends, and acquaintances. Mega thanks to the Inner Traditions team for their expertise and professionalism, especially Albo and Jill for your huge input. You gave the book a coherence it was lacking. And gratitude to Virginia for her stunning map work.

THANK YOU ALL, but especially Jerm!

Index

About the Author

Numerous avatars across the world—some sublime, others more mundane—saw Christopher Freeland start out as an officer in the British Gurkhas in the late 1960s, travel overland to Zambia where he worked as a guide on photographic safaris on foot, and then meet his preceptor and become an orthodox Hindu monk for four years. Those years formed the basis for his real education consisting of in-depth studies of Vedanta, Sanskrit, Ayurveda, and Indian culture. After periods spent in the United States, Europe, Asia, and the Far East, he settled in France and worked as a technical French-English translator, compiling three dictionaries before globalization threw the whole translation world into disarray. Christopher then returned to the study of geomancy, radiesthesia, and Chinese medicine. An eleven-year stint in Thailand resulted in his becoming a full-time radiesthesist, actively applying the lessons learned along the way to resolving geopathic stress, providing solutions for health, researching concerns both natural and man-made, and other related matters. He now alternates between Ireland and Thailand, out in the countryside, running workshops, writing, and offering his devices and services to those in need.

His previous self-publications include the following:

Sanskrit-English Philosophical Wordlist, 1977
Dictionnaire pratique des mondes de la finance et de la bourse, 1996
The Way of the Skeptic: A Blueprint of Concepts from Far and Near for the Serious Seeker of Harmony, 2017
Embracing the Sublime (autobiography), 2019
Radiesthesia I: Method & Training for the Modern Dowser, 2019
Radiesthesia II: Vital Force in the Human Body, 2019
Radiesthesia III: Senses, 2020
Radiesthesia IV: Geopathic Stress, 2020
Other-Dimensional Entities: Their Recognition and Release, 2022

Courses on his method of radiesthesia and its applications are available on radiesthesia.online, along with a variety of other services and products.

BOOKS OF RELATED INTEREST

The Mystery of Doggerland
Atlantis in the North Sea
By Graham Phillips

The Mystery of Skara Brae
Neolithic Scotland and the Origins of Ancient Egypt
by Laird Scranton

The Stones of Time
Calendars, Sundials, and Stone Chambers of Ancient Ireland
by Martin Brennan

Rediscovering Turtle Island
A First Peoples' Account of the Sacred Geography of America
by Taylor Keen

Karahan Tepe
Civilization of the Anunnaki and the
Cosmic Origins of the Serpent of Eden
by Andrew Collins
Foreword by Hugh Newman

Gobekli Tepe: Genesis of the Gods
The Temple of the Watchers and the Discovery of Eden
by Andrew Collins
Introduction by Graham Hancock

Forgotten Civilization
New Discoveries on the Solar-Induced Dark Age
by Robert M. Schoch, Ph.D.
with Catherine Ulissey

Sorcerers of Stone
Architects of the Three Ages
by Camille M. Sauvé

INNER TRADITIONS • BEAR & COMPANY
P.O. Box 388
Rochester, VT 05767
1-800-246-8648
www.InnerTraditions.com
Or contact your local bookseller